Introduction to Botany

Introduction to Botany

Melody Sawyer

R CALLISTO REFERENCE

www.callistoreference.com

Callisto Reference,
118-35 Queens Blvd., Suite 400,
Forest Hills, NY 11375, USA

Visit us on the World Wide Web at:
www.callistoreference.com

ISBN: 978-1-64116-523-5 (Hardback)

Cataloging-in-Publication Data

Introduction to botany / Melody Sawyer.
    p. cm.
Includes bibliographical references and index.
ISBN 978-1-64116-523-5
1. Botany. 2. Plants. 3. Biology. I. Sawyer, Melody.
QK45.2 .I58 2022
580--dc23

TABLE OF CONTENTS

Preface VII

Chapter 1 **Introduction** 1

- Botany   1
- Scope and Importance of Botany   2

Chapter 2 **Branches of Botany** 6

- Paleobotany   6
- Ethnobotany   10
- Plant Ecology   16
- Plant Genetics   23
- Bryology   29
- Pomology   31

Chapter 3 **Plant Cell Biology** 33

- Plant Cells   33
- Types of Plant Cells   36
- Leaf Cells   37
- Plant Cell Organelles   43

Chapter 4 **Plant Tissues** 46

- Vascular Tissue   48
- Epidermal Tissue   65
- Ground Tissue   66
- Meristem   73

Chapter 5 **Plant Anatomy and Physiology** 82

- Plant Anatomy   82
- Plant Physiology   85

- Plant Organ System   95
- The Shoot System   95
- The Root System   160

Permissions

Index

PREFACE

The purpose of this book is to help students understand the fundamental concepts of this discipline. It is designed to motivate students to learn and prosper. I am grateful for the support of my colleagues. I would also like to acknowledge the encouragement of my family.

Botany is a biological field which focuses on the study of plant life. It makes use of diverse techniques and methods such as optical microscopy, live-cell imaging, analysis of chromosome number, electron microscopy, plant chemistry, and function of enzymes and other proteins. Some of the areas of research within this discipline are plant structure, differentiation and growth, reproduction, primary metabolism, and biochemistry. It is also concerned with the study of chemical products, plant development, plant evolutionary relationships, plant diseases and plant taxonomy. The classification of plants within this field is done using various techniques of molecular genetic analysis such as proteomics, genomics and DNA sequences. This textbook is a compilation of chapters that discuss the most vital concepts in the field of botany. It is compiled in such a manner, that it will provide in-depth knowledge about the theory and concepts of this field. For someone with an interest and eye for detail, this book covers the most significant topics related to this field.

A foreword for all the chapters is provided below:

Chapter – Introduction

The branch of biology that studies plant life is referred known as botany. It studies different types of land plants such as vascular plants and bryophytes. This is an introductory chapter which will provide a brief introduction to the significant aspects of botany such as its scope and importance.

Chapter – Branches of Botany

There are various branches of botany such as paleobotany, ethnobotany, plant ecology, plant genetics, bryology and pomology. This chapter closely examines the key concepts of these branches of botany to provide an extensive understanding of the subject.

Chapter – Plant Cell Biology

Plants cells are the eukaryotic cells which are present in green plants. There are different types of plant cells including parenchyma, collenchyma, sclerenchyma, etc. The chapter closely examines the key concepts related to plant cell biology such as these types of plant cells and the different organelles within them to provide an extensive understanding of the subject.

Chapter – Plant Tissues

The combination of similar cells and their extracellular matrix which carries out a specific function is known as a tissue. The different types of plant tissues include vascular tissues, epidermal tissues and ground tissues. The chapter closely examines these types of plant tissues to provide an extensive understanding of the subject.

Chapter – Plant Anatomy and Plant Physiology

The study of the internal structure of plants is referred to as plant anatomy. The sub-discipline of botany that is concerned with the functioning and physiology of plants is known as plant physiology. This chapter has been carefully written to provide an easy understanding of plant anatomy and plant physiology.

Melody Sawyer

Introduction

The branch of biology that studies plant life is known as botany. It studies different types of land plants such as vascular plants and bryophytes. This is an introductory chapter which will provide a brief introduction to the significant aspects of botany such as its scope and importance.

BOTANY

Botany is the scientific study of plants, or multicellular organisms, that carry on photosynthesis. As a branch of biology, botany sometimes is referred to as plant science or plant biology. Botany includes a wide range of scientific subdisciplines that study the structure, growth, reproduction, metabolism, development, diseases, ecology and evolution of plants. The study of plants is important because they are a fundamental part of life on Earth, generating food, oxygen, fuel, medicine and fibers that allow other life forms to exist. Through photosynthesis they absorb carbon dioxide, a waste product generated by most animals and a greenhouse gas that contributes to global warming.

As with other forms of life, plants can be studied at many different levels. One is the molecular level, which is concerned with the biochemical, molecular and genetic functions of plants. Another is the cellular, tissue and organelle (a discrete structure of a cell that has a specialized function) level, which studies the anatomy and physiology of plants; and the community and population level, which involves interactions within a species, with other species and with the environment.

Historically, botanists studied any living being that was not an animal. Although fungi, algae and bacteria now are members of other kingdoms, according to the currently accepted classification system, they usually still are studied in introductory botany classes.

The ancient Greeks were among the first to write about plants in a scientific way. In the fifth century B.C.E., Empedocles believed plants not only had a soul, like animals, but also had reason and common sense. Aristotle believed plants ranked between animals and inanimate objects. Aristotle's pupil Theophrastus wrote two books about plants that still were in use in the 15th century. The Swedish physician-turned-botanist Carl Linné is considered the father of the systematic naming system (nomenclature), which he invented in the 18th century and still is used to give scientific names to all species, plant and otherwise.

Plants always have been convenient organisms to study scientifically because they did not pose the same ethical dilemmas as the study of animals or humans. The Austrian monk Gregor Mendel wrote the first laws of inheritance, a set of primary tenets relating to the transmission of hereditary characteristics from parent organisms to their children, in the 1850s after crossing pea plants in his garden.

SCOPE AND IMPORTANCE OF BOTANY

Hibiscus.

As with other life forms, plant life can be studied from different perspectives, from the molecular, genetic and biochemical level through organelles, cells, tissues, organs, individuals, plant populations, communities of plants, and entire ecosystems. At each of these levels a botanist might be concerned with the classification (taxonomy), structure (anatomy), or function (physiology) of plant life.

Historically, botany covered all organisms that were not considered to be animals. Some of these "plant-like" organisms include fungi (studied in mycology), bacteria and viruses (studied in microbiology), and algae (studied in phycology). Most algae, fungi, and microbes are no longer considered to be in the plant kingdom. However, attention is still given to them by botanists, and bacteria, fungi, and algae are usually covered in introductory botany courses.

Plants are a fundamental part of life on earth. They generate the oxygen, food, fibers, fuel, and medicine that allow higher life forms to exist. Plants also absorb carbon dioxide, a significant greenhouse gas, through photosynthesis. A good understanding of plants is crucial to the future of human societies as it allows us to:

- Feed the world.
- Understand fundamental life processes.

- Utilize medicine and materials.

- Understand environmental changes.

- Maintain ecological, biodiversity, and ecosystem function.

Feed the World

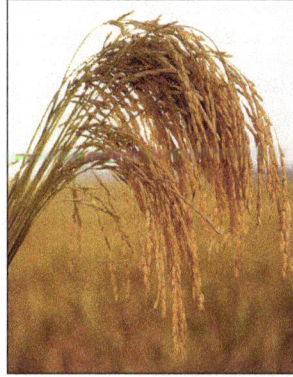

Nearly all the food we eat comes (directly and indirectly)
from plants like this American long grain rice.

Virtually all of the food we eat comes from plants, either directly from staple foods and other fruit and vegetables, or indirectly through livestock, which rely on plants for fodder. In other words, plants are at the base of nearly all food chains, or what ecologists call the first trophic level. Understanding how plants produce the food we eat is therefore important to be able to feed the world and provide food security for future generations, for example through plant breeding.

Not all plants are beneficial to humans, some weeds are a considerable problem in agriculture, and botany provides some of the basic science in order to understand how to minimize their impact. However, other weeds are pioneer plants, which start an abused environment back on the road to rehabilitation, underlining that the term "weed" is a very relative concept, and that broadly defined a weed is simply an undesirable plant that is too successful.

Gregor Mendel laid the foundations of genetics
from his studies of plants.

Understand Fundamental Life Processes

Plants are convenient organisms in which fundamental life processes (like cell division and protein synthesis for example) can be studied, without the ethical dilemmas of studying animals or humans. The genetic laws of inheritance were discovered in this way by Gregor Mendel, who was studying the way pea shape is inherited. What Mendel learned from studying plants has had far reaching benefits outside of botany.

Barbara McClintock discovered transposons, or "jumping genes," by studying maize . These transposons, genes that move from one location to the next on a chromosome, are responsible for the mottled look of maize grains. This sort of research has paved the way for the study of other plant genomes and genome evolution.

Other types of physiological research, including the uptake of carbon by plants through photosynthesis and understanding the physiology behind C3 versus C4 photosynthetic plants, are important for understanding the response of plants to climate change and the feedback mechanisms that occur with increased greenhouse gases in the atmosphere.

These are a few examples that demonstrate how botanical research has an ongoing relevance to the understanding of fundamental biological processes.

Utilize Medicine and Materials

Many of our medicine and recreational drugs, like caffeine and nicotine come directly from the plant kingdom. Aspirin, which originally came from the bark of willow trees, is just one example of a plant derivative used in modern medicine. Pharmacognosy is the study of medicinal and toxic plant derivatives. There may be many novel cures for diseases provided by plants that have not yet been discovered. Popular stimulants like coffee, chocolate, tobacco, and tea also come from plants. Most alcoholic beverages come from fermenting plants such as barley malt and grapes.

Plants also provide us with many natural materials, such as cotton, wood, paper, linen, vegetable oils, some types of rope, and rubber. The production of silk would not be possible without the cultivation of the mulberry plant. Sugarcane and other plants have recently been put to use as sources of biofuels, which are important alternatives to fossil fuels. Plants are extremely valuable as recreation for millions of people who enjoy gardening, horticultural, and culinary uses of plants every day.

Understand Environmental Changes

Plants can also help us understand changes in on our environment in many ways:

- Understanding habitat destruction and species extinction is dependent on an accurate and complete catalog of plant systematics and taxonomy.

- Plant responses to ultraviolet radiation can help us monitor problems like ozone depletion.

- Analyzing pollen deposited by plants thousands or millions of years ago can help scientists to reconstruct past climates and predict future ones, an essential part of climate change research.

- Recording and analyzing the timing of plant life cycles are important parts of phenology used in climate-change research.

- Plants can act a bit like the 'miner's canary', an early warning system, alerting us to important changes in our environment. For example, lichen, which are sensitive to atmospheric conditions, have been extensively used as pollution indicators.

Branches of Botany

There are various branches of botany such as paleobotany, ethnobotany, plant ecology, plant genetics, bryology and pomology. This chapter closely examines the key concepts of these branches of botany to provide an extensive understanding of the subject.

PALEOBOTANY

Paleobotany is the branch of paleontology dealing with the recovery and identification of plant remains from geological contexts, and their use in the reconstruction of past environments and the history of life.

The parent discipline, paleontology, is the study of the developing history of life on Earth based on the fossil record, with paleobotany dealing with plant remains, paleozoology with animal remains, and micropaleontology with microfossils. Paleobotany includes the study of terrestrial plant fossils as well as the study of marine autotrophs, such as algae. A closely related field to paleobotany is palynology, the study of fossil and extant spores and pollen.

Paleobotany not only addresses the inner nature of humans to know more about the history of life, but also has practical application today, helping people to better understand such aspects as climate change.

Paleobotany is important in the reconstruction of prehistoric ecological systems and climate, known as paleoecology and paleoclimatology respectively, and is fundamental to the study of plant development and evolution. Paleobotany has also become important to the field of archaeology, primarily for the use of phytoliths ("plant stone") in relative dating and in paleoethnobotany. Paleobotany shows one of the core values of science, that is, the willingness of the scientific community to work beyond borders of particular disciplines.

Paleobotanical Record

Macroscopic remains of true vascular plants are first found in the fossil record during the Silurian period. Some dispersed, fragmentary fossils of disputed affinity, primarily spores and cuticles, have been found in rocks from the Ordovician period of Oman and are thought to derive from liverwort- or moss-grade fossil plants.

An important early land plant fossil locality is the Rhynie Chert, an Early Devonian sinter (hot spring) deposit composed primarily of silica found outside the town of Rhynie in Scotland. An unpolished hand sample of the Lower Devonian Rhynie Chert from Scotland.

The Rhynie Chert is exceptional due to its preservation of several different clades of plants, from mosses and lycopods to more unusual, problematic forms. Many fossil animals, including arthropods and arachnids, are also found in the Rhynie Chert, and it offers a unique window on the history of early terrestrial life.

Plant-derived macrofossils become abundant in the Late Devonian and include tree trunks, fronds, and roots. The earliest known tree is Archaeopteris, which bears simple, fern-like leaves spirally arranged on branches atop a conifer-like trunk.

Widespread coal swamp deposits across North America and Europe during the Carboniferous period contain a wealth of fossils containing arborescent lycopods up to 30 meters tall, abundant seed plants, such as conifers and seed ferns, and countless smaller, herbaceous plants.

Angiosperms (flowering plants) evolved during the Mesozoic, and flowering plant pollen and leaves first appear during the Early Cretaceous, approximately 130 million years ago.

Palynology

Palynology is the science that studies contemporary and fossil palynomorphs; that is, particles of a size between five and 500 micrometres, found in rock deposits, and composed of organic material. Such palynomorphs studied include pollen, spores, dinoflagellate cysts, acritarchs, chitinozoans, and scolecodonts, together with particulate organic matter (POM) and kerogen found in sedimentary rocks and sediments.

Palynology is a branch of earth science (geology or geological science) and biological science (biology), particularly plant science (botany). Stratigraphical palynology is a branch of micropalaeontology and paleobotany that studies fossil palynomorphs from the Precambrian to the Holocene.

The term palynology was introduced by Hyde and Williams in 1944, following correspondence with the Swedish geologist Antevs, in the pages of the Pollen Analysis Circular (one of the first journals devoted to pollen analysis, and produced by Paul Sears in North America). Hyde and Williams chose palynology on the basis of the Greek words paluno meaning to sprinkle, and pale meaning dust (and thus similar to the Latin word pollen).

Methods of Study

Palynomorphs are broadly defined as organic-walled microfossils between five and 500 micrometers in size. They are extracted from rocks and sediments both physically, by wet sieving, often after ultrasonic treatment, and chemically, by using chemical digestion to remove the non-organic fraction. For example, palynomorphs may be extracted using hydrochloric acid (HCl) to digest carbonate minerals, and hydrofluoric acid (HF) to digest silicate minerals in suitable fume cupboards in specialist laboratories.

Samples are then mounted on microscope slides and examined using light microscopy or scanning electron microscopy. Once the pollen grains have been identified they can be plotted on a pollen diagram that is then used for interpretation. Pollen diagrams are useful in giving evidence of past human activity (anthropogenic impact), vegetation history, and climatic history.

Palynology uses many techniques from other related fields such as geology, botany, paleontology, archaeology, pedology (soil study), and geography.

Applications

Palynology is used for a diverse range of applications, related to many scientific disciplines:

- Biostratigraphy and geochronology: Geologists use palynological studies in biostratigraphy to correlate strata and determine the relative age of a given bed, horizon, formation, or stratigraphical sequence.

- Paleoecology and climate change: Palynology can be used to reconstruct past vegetation (land plants) and marine and freshwater phytoplankton communities, and so infer past environmental (paleoenvironmental) and paleoclimatic conditions.

- Organic palynofacies studies: These studies examine the preservation of the particulate organic matter and palynomorphs, and provide information on the depositional environment of sediments and depositional palaeoenvironments of sedimentary rocks.

- Geothermal alteration studies: These studies examine the color of palynomorphs extracted from rocks to give the thermal alteration and maturation of sedimentary sequences, which provides estimates of maximum paleotemperatures.

- Limnology studies: Freshwater polynomorphs and animal and plant fragments, including the prasinophytes and desmids (green algae) can be used to study past lake levels and long-term climate change.

- Taxonomy and evolutionary studies.

- Forensic palynology: Forensic palynology is the study of pollen and other palynomorphs for evidence at a crime scene.

- Allergy studies: Studies of the geographic distribution and seasonal production of pollen, can help sufferers of allergies such as hay fever.

- Melissopalynology: This is the study of pollen and spores found in honey.

Because the distribution of acritarchs, chitinozoans, dinoflagellate cysts, pollen, and spores provides evidence of stratigraphical correlation through biostratigraphy and paleoenvironmental reconstruction, one common and lucrative application of palynology is in oil and gas exploration.

Palynology also allows scientists to infer the climatic conditions from the vegetation present in an area thousands or millions of years ago. This is a fundamental part of research into climate change.

Paleoecology

Paleoecology uses data from fossils and subfossils to reconstruct the ecosystems of the past. It includes the study of fossil organisms in terms of their life cycle, their living interactions, their natural environment, their manner of death, and their burial.

Paleoecology's aim is therefore to build the most detailed model possible of the life environment of those living organisms that are found today as fossils; such reconstruction work involves complex interactions among environmental factors (temperature, food supplies, degree of solar illumination, etc.). Of course, much of this complex data has been distorted or destroyed by the post-mortem fossilization processes, adding another layer of complexity.

The environmental complexity factor is normally tackled through statistical analysis of the available numerical data (quantitative paleontology or paleostatistics), while post-mortem processes as a source of information are known as the field of taphonomy.

Much paleoecological research focuses on the last two million years (formerly known as the Quaternary period), because older environments are less well-represented in the fossil timeline of evolution. Indeed, many studies concentrate on the Holocene epoch (the last 10,000 years), or the last glacial stage of the Pleistocene epoch (the Wisconsin/ Weichsel/Devensian/Würm glaciation]] of the ice age, from 50,000 to 10,000 years ago). Such studies are useful for understanding the dynamics of ecosystem change and for reconstructing pre-industrialization ecosystems. Many public policy decision makers have pointed to the importance of using paleoecological studies as a basis for choices made in conservation ecology. Often paleoecologists will use cores from lakes or bogs to reconstruct pollen assemblages, lithology, and to perform geochemical analysis. These tools aid in determining the species composition and climatic conditions, which can contribute to the understanding of how ecosystems change and have changed with climatic and environmental conditions.

ETHNOBOTANY

Ethnobotany is the systematic study of the relationships between plants and people. It is not simply the study of the human "use" of plants; rather, ethnobotany locates plants within their cultural context in particular societies, and situates peoples within their ecological contexts. Ethnobotanists examine:

- The culturally specific ways that humans perceive and classify different kinds of plants.

- The things humans do to plant species, such as destroying "weeds" or "domesticating" and planting specific kinds of food and medicinal plants.

- The ways in which various members of the plant world influence human cultures.

This inquiry ranges from the geopolitical impact of the European demand for spices (which helped to launch the Age of Exploration) to the role of hallucinogenic snuffs used by Amazonian shamans in religious rituals.

Attributes such as creativity, reason, and curiosity, coupled with a desire to benefit others—attributes common in the scientific community—aids those studying ethnobotany to make important contributions. For example, the study of indigenous food production and local medicinal knowledge offers the promise of practical implications for developing sustainable agriculture and discovering new medicines.

The term "ethnobotany" was coined in 1895, by J.M. Harshberger, an American botanist at the University of Pennsylvania. Modern ethnobotany is an interdisciplinary field drawing together scholars from anthropology, botany, archaeology, geography, medicine, linguistics, economics, landscape architecture, and pharmacology.

Ethnobotany is considered a branch of ethnobiology, the study of past and present interrelationships between human cultures and the plants, animals, and other organisms in their environment. Like its parent field, ethnobotany makes apparent the connection between human cultural practices and the sub-disciplines of biology.

Ethnobotanical studies range across space and time, from archaeological investigations of the role of plants in ancient civilizations to the bioengineering of new crops. Furthermore, ethnobotany is not limited to nonindustrialized or nonurbanized societies. In fact, co-adaptation of plants and human cultures has changed—and perhaps intensified—in the context of urbanization and globalization in the twentieth and twenty-first centuries. Nonetheless, indigenous, non-Westernized cultures play a crucial role in ethnobotany, as they possess a previously undervalued knowledge of local ecology gained through centuries or even millennia of interaction with their biotic (living) environment.

The significance of ethnobotany is manifold. The study of indigenous food production and local medicinal knowledge may have practical implications for developing sustainable agriculture and discovering new medicines. Ethnobotany also encourages an awareness of the link between biodiversity and cultural diversity, as well as a sophisticated understanding of the mutual influence (both beneficial and destructive) of plants and humans.

The Influence of Plants on Human Culture

Why might plants have come to function as the material basis for human culture? The combination of their immobility (terrestrial plants must remain rooted in the soil) and tremendous production of cellulose makes plants a far more efficient and reliable source of building materials and food than animals.

The biochemical diversity of plants, which contributes to their myriad medicinal and dietary uses, might also be traced in part to their immobility. Plants produce chemicals as a way of interacting with other organisms in their environment, either for mutual gain— such as enlisting animals in the transport of pollen or seeds—or as a mechanism of defense, to repel or poison predators or parasites. Modern societies depend on chemical agents in plants for 25 percent of prescription drugs and nearly all recreational chemicals, such as the caffeine in coffee, the nicotine in tobacco, and the theophylline in tea.

Historic Roots of Ethnobotany

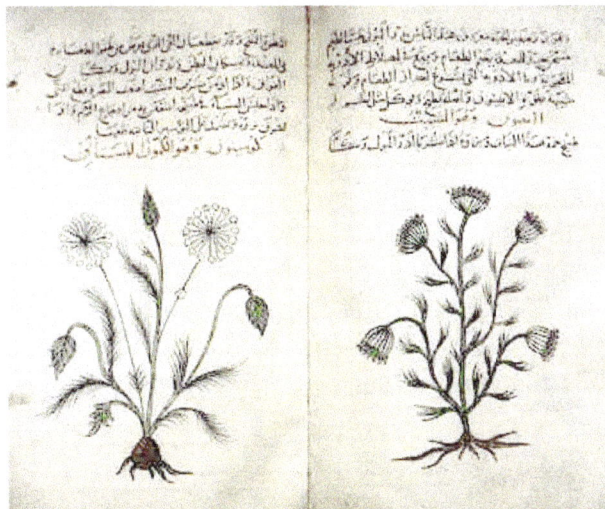

An Arabic edition of Dioscorides's De Materia Medica (circa 1334) describes the medicinal features of cumin and dill.

Although ethnobotany did not emerge as an academic discipline until the end of the nineteenth century, its roots extend back to Greek, Roman, and Islamic sources. In 77 C.E., the Greek surgeon Dioscorides published De Materia Medica, a catalog of about 600 plants found in the Mediterranean. This illustrated book of herbal (a book that describes the appearance, medicinal properties, and other characteristics of plants used in herbal medicine), which influenced scholars through the Middle Ages, contained information on how and when each plant was gathered, its use by the Greeks, and whether or not it was edible. (Dioscorides even provided recipes.) He also assessed the economic potential of these plants.

However, the systematic study of plants was not confined to the West: The earliest known herbal was compiled by Chinese emperor Shen Nung sometime before 2000 B.C.E., and both the Incas of South America and the Aztecs of Mesoamerica maintained botanical gardens.

The Renaissance in Europe saw a revival of interest in ethnobotany, which was intensified by geographic exploration and later colonialism. In 1542, Renaissance artist Leonhart Fuchs published De Historia Stirpium, a catalogue of 400 plants native to Germany and Austria. John Gerard (1545-1611/12) published the most popular of sixteenth century herbals, the General Historie of Plants, which remained in print for over 400 years. John Ray provided the first definition of species in his Historia Plantarum.

In 1753, the Swedish botanist Carl Linnaeus wrote Species Plantarum, which included information on approximately 5,900 plants. Linnaeus, known as "the father of taxonomy," is famous for popularizing the binomial method of nomenclature, in which all living organisms are assigned a two-part name (genus, species).

The nineteenth century saw the peak of botanical exploration. Alexander von Humboldt collected data from "the New World," and the famous Captain Cook brought back information on plants from the South Pacific. At this time, major botanical gardens were founded in Europe, such as the Royal Botanic Gardens, Kew (commonly known as Kew Gardens).

The modern discipline of ethnobotany began to emerge in the late nineteenth century in part out of field-work concentrated in the north American West. Researchers referred to their work as "aboriginal botany," which studied the forms of plant-life used by aboriginal peoples. From the 1860s to the 1890s, Edward Palmer collected artifacts and botanical specimens from peoples in the Great Basin region and Mexico. Other scholars who analyzed the uses of plants under an indigenous/local perspective included Matilda Coxe Stevenson, Zuni plants; Frank Cushing, Zuni foods; and the team of Wilfred Robbins, J.P. Harrington, and Barbara Freire-Marreco, Tewa pueblo plants.

Modern Ethnobotany

Beginning in the twentieth century, the field of ethnobotany experienced a shift from the raw compilation of data to a greater methodological and conceptual reorientation.

Today, the practice of ethnobotany requires a variety of skills:

- Botanical training for the identification and preservation of plant specimens.

- Anthropological training to understand the cultural concepts around the perception of plants.

- Linguistic training to transcribe local terms and understand native morphology, syntax, and semantics.

Ethnobotanists engage in a broad array of research questions and practices, which do not lend themselves to easy categorization. However, the following headings attempt to describe some of the key areas of modern study.

Ethnomedicine

The drug atropine has its origins in Atropa belladonna
(or "deadly nightshade").

Ethnomedicine is a sub-field of medical anthropology that deals with the study of traditional medicines—not only those with relevant written sources (e.g., Traditional Chinese Medicine and Ayurveda), but also those whose knowledge and practices have been orally transmitted over the centuries.

While the focus of ethnomedical studies is often the indigenous perception and use of traditional medicines, another stimulus for this type of research is drug discovery and development. Major pharmaceuticals such as digoxin, morphine, and atropine have been traced to foxglove, opium, and belladonna, respectively. Ethnomedical investigations in this century have led to the development of important drugs such as reserpine (a treatment for hypertension) podophyllotoxin (the base of an important anti-cancer drug), and vinblastine (used in the treatment of certain cancers).

Agriculture

Agriculture may be defined as the culturally influenced selection of plants with specific genetic characteristics that are desired by humans to create domesticated plants, or crops.

Ethnobotany contributes to an understanding of agriculture in two ways:

- By revealing ways to create genetically altered plants for human purposes.

- By describing and explaining the many different ways the same crop can be raised, whether for economic gain, a desire for sustained yield, or other culturally specific purposes.

One example of the mutual influence of plants and human cultures is illustrated by the Irish potato famine of the mid-nineteenth century. The Irish cultivation of potatoes was an example of monoculture, the practice of planting crops with the same patterns of growth resulting from genetic similarity. Monoculture can lead to large scale crop failure when the single genetic variant (or cultivar) becomes susceptible to a disease. The famine, which resulted in somewhere between 500,000 and one million deaths, was caused by the cultivar's susceptibility to Phytophthora infestans. The famine partially triggered widespread Irish immigration to Great Britain, the United States, Canada, and Australia.

Plants in Religion and Ritual

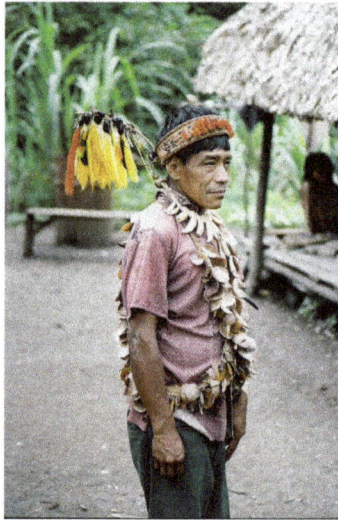

An Urarina shaman.

An entheogen, in the strictest sense, is a psychoactive substance (most often some plant matter with hallucinogenic effects) that occasions an enlightening spiritual or mystical experience. Entheogens have played a pivotal role in the spiritual practices of most American cultures for millennia. One of the founders of modern ethnobotany, Richard Evans Schultes of Harvard University, documented the ritual use of peyote cactus among the Kiowa who live in what has become Oklahoma in the United States. Used traditionally by many cultures of what is now Mexico, peyote spread to North America in the nineteenth century, replacing the toxic entheogen Sophora secundiflora (mescal bean).

Indigenous peoples of South America employ a wide variety of entheogens. Better-known examples include ayahuasca (Banisteriopsis caapi plus admixtures) among indigenous peoples (such as the Urarina) of Peruvian Amazonia. Other well-known entheogens include: Borrachero (Brugmansia spp); San Pedro (Trichocereus spp); and various tryptamine-bearing snuffs. The familiar tobacco plant, when used uncured in large doses in shamanic contexts, also serves as an entheogen in South America.

Folk Classification

Folk classification refers to how members of a language community name and categorize plants and animals. This type of ethnobotanical study relies on an emic approach: That is, a description of behavior in terms meaningful (consciously or unconsciously) to the actor.

The first individual to study an emic perspective of the plant world was Leopold Glueck, a German physician working in Sarajevo. His 1896 publication on the traditional medicinal uses of plants by rural people in Bosnia may be considered the first modern ethnobotanical work.

Archaeoethnobotany

Guila Naquitz Cave, site of the oldest known remains of maize.

Archaeoethnobotany (or paleoethnobotany) is the study of the ethnobotany of the ancient past. It is closely linked to ethnobotany, as it is difficult to understand the ecology of modern environments without considering the environmental history that often involves prehistoric human interventions.

The history of the domestication of the cereal grain maize (commonly known as "corn") is of particular interest to archaeoethnobotanists. The process is thought by some to have started 7,500 to 12,000 years ago. Recent genetic evidence suggests that maize domestication occurred 9000 years ago in central Mexico, perhaps in the highlands between Oaxaca and Jalisco. Archaeological remains of early maize cobs, found at Guila

Naquitz Cave in the Oaxaca Valley, date back roughly 6,250 years; the oldest cobs from caves near Tehuacan, Puebla, have been dated to approximately 2750 B.C.E.

PLANT ECOLOGY

Plant ecology is a subdiscipline of ecology which studies the distribution and abundance of plants, the effects of environmental factors upon the abundance of plants, and the interactions among and between plants and other organisms. Examples of these are the distribution of temperate deciduous forests in North America, the effects of drought or flooding upon plant survival, and competition among desert plants for water, or effects of herds of grazing animals upon the composition of grasslands.

A tropical plant community on Diego Garcia.

Rangeland monitoring using Parker 3-step Method.

A global overview of the Earth's major vegetation types is provided by O.W. Archibold. He recognizes 11 major vegetation types: tropical forests, tropical savannas, arid regions (deserts), Mediterranean ecosystems, temperate forest ecosystems, temperate grasslands, coniferous forests, tundra (both polar and high mountain), terrestrial wetlands, freshwater ecosystems and coastal/marine systems. This breadth of topics shows the complexity of plant ecology, since it includes plants from floating single-celled algae up to large canopy forming trees.

One feature that defines plants is photosynthesis. Photosynthesis is the process of a chemical reactions to create glucose and oxgyen, which is vital for plant life. One of the most important aspects of plant ecology is the role plants have played in creating the oxygenated atmosphere of earth, an event that occurred some 2 billion years ago. It can be dated by the deposition of banded iron formations, distinctive sedimentary rocks with large amounts of iron oxide. At the same time, plants began removing carbon dioxide from the atmosphere, thereby initiating the process of controlling Earth's climate. A long term trend of the Earth has been toward increasing oxygen and decreasing carbon dioxide, and many other events in the Earth's history, like the first movement of life onto land, are likely tied to this sequence of events.

One of the early classic books on plant ecology was written by J.E. Weaver and F.E. Clements. It talks broadly about plant communities, and particularly the importance of forces like competition and processes like succession.

Plant ecology can also be divided by levels of organization including plant ecophysiology, plant population ecology, community ecology, ecosystem ecology, landscape ecology and biosphere ecology.

The study of plants and vegetation is complicated by their form. First, most plants are rooted in the soil, which makes it difficult to observe and measure nutrient uptake and species interactions. Second, plants often reproduce vegetatively, that is asexually, in a way that makes it difficult to distinguish individual plants. Indeed, the very concept of an individual is doubtful, since even a tree may be regarded as a large collection of linked meristems. Hence, plant ecology and animal ecology have different styles of approach to problems that involve processes like reproduction, dispersal and mutualism. Some plant ecologists have placed considerable emphasis upon trying to treat plant populations as if they were animal populations, focusing on population ecology. Many other ecologists believe that while it is useful to draw upon population ecology to solve certain scientific problems, plants demand that ecologists work with multiple perspectives, appropriate to the problem, the scale and the situation.

Plant ecology has its origin in the application of plant physiology to the questions raised by plant geographers. Carl Ludwig Willdenow was one of the first to note that similar climates produced similar types of vegetation, even when they were located in different parts of the world. Willdenow's student, Alexander von Humboldt, used physiognomy to describe vegetation types and observed that the distribution vegetation types was based on environmental factors. Later plant geographers who built upon Humboldt's work included Joakim Frederik Schouw, A.P. de Candolle, August

Grisebach and Anton Kerner von Marilaun. Schouw's work, published in 1822, linked plant distributions to environmental factors (especially temperature) and established the practice of naming plant associations by adding the suffix *-etum* to the name of the dominant species.

Alexander von Humboldt's work connecting plant distributions with environmental factors played an important role in the genesis of the discipline of plant ecology.

Working from herbarium collections, De Candolle searched for general rules of plant distribution and settled on using temperature as well. Grisebach's two-volume work, *Die Vegetation der Erde nach Ihrer Klimatischen Anordnung*, published in 1872, saw plant geography reach its "ultimate form" as a descriptive field.

Starting in the 1870s, Swiss botanist Simon Schwendener, together with his students and colleagues, established the link between plant morphology and physiological adaptations, laying the groundwork for the first ecology textbooks, Eugenius Warming's Plantesamfund and Andreas Schimper's 1898 Pflanzengeographie auf Physiologischer Grundlage. Warming successfully incorporated plant morphology, physiology taxonomy and biogeography into plant geography to create the field of plant ecology. Although more morphological than physiological, Schimper's has been considered the beginning of plant physiological ecology. Plant ecology was initially built around static ideas of plant distribution; incorporating the concept of succession added an element to change through time to the field. Henry Chandler Cowles' studies of plant succession on the Lake Michigan sand dunes and Frederic Clements' 1916 monograph on the subject established it as a key element of plant ecology.

Plant ecology developed within the wider discipline of ecology over the twentieth century. Inspired by Warming's *Plantesamfund*, Arthur Tansley set out to map British plant communities. In 1904 he teamed up with William Gardner Smith and others involved in vegetation mapping to establish the Central Committee for the Survey and Study of British Vegetation, later shortened to British Vegetation Committee. In 1913, the British Vegetation Committee organised the British Ecological Society (BES), the first professional society of ecologists. This was followed in 1917 by the establishment of the Ecological Society of America (ESA); plant ecologists formed the largest subgroup among the inaugural members of the ESA.

Cowles' students played an important role in the development of the field of plant ecology during the first half of the twentieth century, among them William S. Cooper, E. Lucy Braun and Edgar Transeau.

Distribution

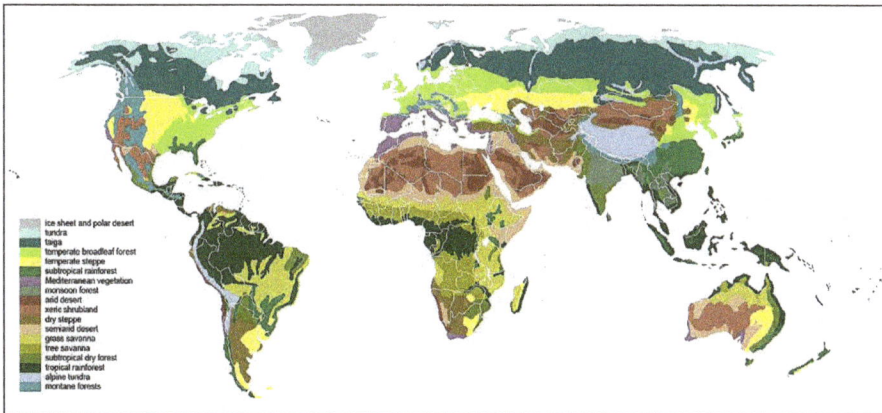

World biomes are based upon the type of dominant plant.

Plant distributions is governed by a combination of historical factors, ecophysiology and biotic interactions. The set of species that can be present at a given site is limited by historical contingency. In order to show up, a species must either have evolved in an area or dispersed there (either naturally or through human agency), and must not have gone locally extinct. The set of species present locally is further limited to those that possess the physiological adaptations to survive the environmental conditions that exist. This group is further shaped through interactions with other species.

Plant communities are broadly distributed into biomes based on the form of the dominant plant species. For example, grasslands are dominated by grasses, while forests are dominated by trees. Biomes are determined by regional climates, mostly temperature and precipitation, and follow general latitudinal trends. Within biomes, there may be many ecological communities, which are impacted not only by climate and a variety of smaller-scale features, including soils, hydrology, and disturbance regime. Biomes also change with elevation, high elevations often resembling those found at higher latitudes.

Biological Interactions

Competition

Plants, like most life forms, require relatively few basic elements: carbon, hydrogen, oxygen, nitrogen, phosphorus and sulphur; hence they are known as CHNOPS life forms. There are also lesser elements needed as well, frequently termed micronutrients, such as magnesium and sodium. When plants grow in close proximity, they may deplete supplies of these elements and have a negative impact upon neighbours. Competition for resources vary from complete symmetric (all individuals receive the same amount of resources, irrespective of their size) to perfectly size symmetric (all individuals exploit the same amount of resource per unit biomass) to absolutely size-asymmetric (the largest individuals exploit all the available resource). The degree of size asymmetry has major effects on the structure and diversity of ecological communities.In many cases (perhaps most) the negative effects upon neighbours arise from size asymmetric competition for light. In other cases, there may be competition below ground for water, nitrogen, or phosphorus. To detect and measure competition, experiments are necessary; these experiments require removing neighbours, and measuring responses in the remaining plants. Many such studies are required before useful generalizations can be drawn.

Overall, it appears that light is the most important resource for which plants compete, and the increase in plant height over evolutionary time likely reflects selection for taller plants to better intercept light. Many plant communities are therefore organized into hierarchies based upon the relative competitive abilities for light. In some systems, particularly infertile or arid systems, below ground competition may be more significant. Along natural gradients of soil fertility, it is likely that the ratio of above ground to below ground competition changes, with higher above ground competition in the more fertile soils. Plants that are relatively weak competitors may escape in time (by surviving as buried seeds) or in space (by dispsersing to a new location away from strong competitors).

In principle, it is possible to examine competition at the level of the limiting resources if a detailed knowledge of the physiological processes of the competing plants is available. However, in most terrestrial ecological studies, there is only little information on the uptake and dynamics of the resources that limit the growth of different plant species, and, instead, competition is inferred from observed negative effects of neighbouring plants without knowing precisely which resources the plants were competing for. In certain situations, plants may compete for a single growth-limiting resource, perhaps for light in agricultural systems with sufficient water and nutrients, or in dense stands of marsh vegetation, but in many natural ecosystems plants may be colimited by several resources, e.g. light, phosphorus and nitrogen at the same time.

Therefore, there are many details that remain to be uncovered, particularly the kinds of competition that arise in natural plant communities, the specific resource(s), the relative importance of different resources, and the role of other factors like stress or disturbance in regulating the importance of competition.

Mutualism

Mutualism is defined as an interaction "between two species or individuals that is beneficial to both". Probably the most widespread example in plants is the mutual beneficial relationship between plants and fungi, known as mycorrhizae. The plant is assisted with nutrient uptake, while the fungus receives carbohydrates. Some the earliest known fossil plants even have fossil mycorrhizae on their rhizomes.

The flowering plants are a group that have evolved by using two major mutualisms. First, flowers are pollinated by insects. This relationship seems to have its origins in beetles feeding on primitive flowers, eating pollen and also acting (unwittingly) as pollinators. Second, fruits are eaten by animals, and the animals then disperse the seeds. Thus, the flowering plants actually have three major types of mutualism, since most higher plants also have mycorrhizae.

Plants may also have beneficial effects upon one another, but this is less common. Examples might include "nurse plants" whose shade allows young cacti to establish. Most examples of mutualism, however, are largely beneficial to only one of the partners, and may not really be true mutualism. The term used for these more one-sided relationships, which are mostly beneficial to one participant, is facilitation. Facilitation among neighboring plants may act by reducing the negative impacts of a stressful environment. In general, facilitation is more likely to occur in physically stressful environments than in favorable environments, where competition may be the most important interaction among species.

Commensalism is similar to facilitation, in that one plant is mostly exploiting another. A familiar example is the ephiphytes which grow on branches of tropical trees, or even mosses which grow on trees in deciduous forests.

It is important to keep track of the benefits received by each species to determine the appropriate term. Although people are often fascinated by unusual examples, it is important to remember that in plants, the main mutualisms are mycorrhizae, pollination, and seed dispersal.

Herbivory

An important ecological function of plants is that they produce organic compounds for herbivores in the bottom of the food web. A large number of plant traits, from thorns to chemical defenses, can be related to the intensity of herbivory. Large herbivores can also have many effects on vegetation. These include removing selected species, creating gaps for regeneration of new individuals, recycling nutrients, and dispersing seeds. Certain ecosystem types, such as grasslands, may be dominated by the effects of large herbivores, although fire is also an equally important factor in this biome. In few cases, herbivores are capable of nearly removing all the vegetation at a site (for example, geese in the Hudson Bay Lowlands of Canada, and nutria in the marshes of Louisiana) but normally herbivores have a more selective impact, particularly when large predators control the abundance of herbivores. The usual method of studying the effects of herbivores is to build exclosures, where they cannot feed, and compare the plant communities in the exclosures to those outside over many years. Often such long term experiments show that herbivores have a significant effect upon the species that make up the plant community.

Reindeer in front of herbivore exclosures. Excluding different herbivores (here reindeer, or reindeer and rodents) has different effects on the vegetation.

Abundance

The ecological success of a plant species in a specific environment may be quantified by its abundance, and depending on the life form of the plant different measures of abundance may be relevant, e.g. density, biomass, or plant cover.

The change in the abundance of a plant species may be due to both abiotic factors, e.g. climate change, or biotic factors, e.g. herbivory or interspecific competition.

Colonisation and Local Extinction

Whether a plant species is present at a local area depends on the processes of colonisation and local extinction. The probability of colonisation decreases with distance to neighboring habitats where the species is present and increases with plant abundance and fecundity in neighboring habitats and the dispersal distance of the species. The probability of local extinction decreases with abundance (both living plants and seeds in the soil seed bank).

Life Forms

Reproduction

There are a few ways that reproduction occurs within plant life, and one way is through parthenogenesis. Parthenogenesis is defined as "a form of asexual reproduction in which genetically identical offspring (clones) are produced". Another form of reproduction is through cross-fertilization, which is defined as "fertilization in which the egg and sperm are produced by different individuals", and in plants this occurs in the ovule. Once an ovule is fertilized within the plant this becomes what is known as a seed. A seed normally contains the nutritive tissue also known as the endosperm and the embryo. A seedling is a young plant that has recently gone through germination. Another form of reproduction of a plant is self-fertilization; in which both the sperm and the egg are produced from the same individual- this plant is therefore a self-compatible titled plant.

PLANT GENETICS

Plant genetics is the study of genes, genetic variation, and heredity specifically in Plants. It is generally considered a field of biology and botany, but intersects frequently with many other life sciences and is strongly linked with the study of information systems. Plant genetics is similar in many ways to animal genetics but differs in a few key areas.

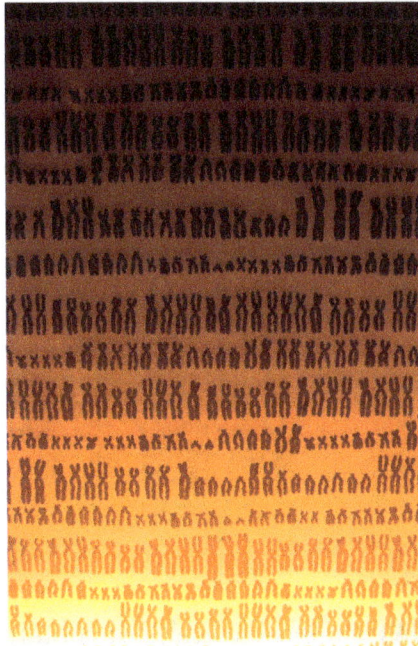

An image of multiple chromosomes, taken from many cells.

The discoverer of genetics was Gregor Mendel, a late 19th-century scientist and Augustinian friar. Mendel studied "trait inheritance", patterns in the way traits are handed down from parents to offspring. He observed that organisms (most famously pea plants) inherit traits by way of discrete "units of inheritance". This term, still used today, is a somewhat ambiguous definition of what is referred to as a gene. Much of Mendel's work with plants still forms the basis for modern plant genetics.

Plants, like all known organisms, use DNA to pass on their traits. Animal genetics often focuses on parentage and lineage, but this can sometimes be difficult in plant genetics due to the fact that plants can, unlike most animals, be self-fertile. Speciation can be easier in many plants due to unique genetic abilities, such as being well adapted to polyploidy. Plants are unique in that they are able to produce energy-dense carbohydrates via photosynthesis, a process which is achieved by use of Chloroplast. Chloroplasts, like the superficially similar mitochondria, possess their own DNA. Chloroplasts thus provide an additional reservoir for genes and genetic diversity, and an extra layer of genetic complexity not found in animals.

The study of plant genetics has major economic impacts: many staple crops are genetically modified to increase yields, confer pest and disease resistance, provide resistance to herbicides, or to increase their nutritional value.

DNA

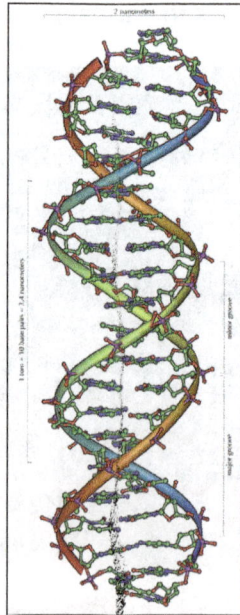

The structure of part of a DNA double helix.

Deoxyribonucleic acid (DNA) is a nucleic acid that contains the genetic instructions used in the development and functioning of all known living organisms and some viruses. The main role of DNA molecules is the long-term storage of information. DNA is often compared to a set of blueprints or a recipe, or a code, since it contains the instructions needed to construct other components of cells, such as proteins and RNA molecules. The DNA segments that carry this genetic information are called genes, and their location within the genome are referred to as genetic loci, but other DNA sequences have structural purposes, or are involved in regulating the use of this genetic information.

Geneticists, including plant geneticists, use this sequence of DNA to their advantage to better find and understand the role of different genes within a given genome. Through research and plant breeding, manipulation of different plant genes and loci encoded by the DNA sequence of the plant chromosomes by various methods can be done to produce different or desired genotypes that result in different or desired phenotypes.

Plant Specific Genetics

Plants, like all other known living organisms, pass on their traits using DNA. Plants however are unique from other living organisms in the fact that they have Chloroplasts. Like mitochondria, chloroplasts have their own DNA. Like animals, plants experience somatic mutations regularly, but these mutations can contribute to the germ line with ease, since flowers develop at the ends of branches composed of somatic cells. People have known of this for centuries, and mutant branches are called "sports". If the fruit on the sport is economically desirable, a new cultivar may be obtained.

Some plant species are capable of self-fertilization, and some are nearly exclusively self-fertilizers. This means that a plant can be both mother and father to its offspring, a rare occurrence in animals. Scientists and hobbyists attempting to make crosses between different plants must take special measures to prevent the plants from self-fertilizing. In plant breeding, people create hybrids between plant species for economic and aesthetic reasons. For example, the yield of Corn has increased nearly five-fold in the past century due in part to the discovery and proliferation of hybrid corn varieties. Plant genetics can be used to predict which combination of plants may produce a plant with Hybrid vigor, or conversely many discoveries in Plant genetics have come from studying the effects of hybridization.

Plants are generally more capable of surviving, and indeed flourishing, as polyploids. Polyploid organisms have more than two sets of homologous chromosomes. For example, humans have two sets of homologous chromosomes, meaning that a typical human will have 2 copies each of 23 different chromosomes, for a total of 46. Wheat on the other hand, while having only 7 distinct chromosomes, is considered a hexaploid and has 6 copies of each chromosome, for a total of 42. In animals, inheritable germline polyploidy is less common, and spontaneous chromosome increases may not even survive past fertilization. In plants however this is no such problem, polyploid individuals are created frequently by a variety of processes, however once created usually cannot cross back to the parental type. Polyploid individuals, if capable of self-fertilizing, can give rise to a new genetically distinct lineage, which can be the start of a new species. This is often called "instant speciation". Polyploids generally have larger fruit, an economically desirable trait, and many human food crops, including wheat, maize, potatoes, peanuts, strawberries and tobacco, are either accidentally or deliberately created polyploids.

Arabidopsis thaliana, growing from between a crack in a sidewalk; it is considered a key model organism in plant genetics.

Model Organisms

Arabidopsis Thaliana

Arabidopsis thaliana, also known as thale cress, has been the model organism for the study of plant genetics. As Drosphila, a species of fruit fly, was to the understanding of early genetics, so has been arabidopsis to the understanding of plant genetics. It was the first plant to ever have its genome sequenced in the year 2000. It has a small genome, making the initial sequencing more attainable. It has a genome size of 125 Mbp that encodes about 25,000 genes. Because an incredible amount of research has been done on the plant, a database called The Arabidopsis Information Resource (TAIR) has been established as a repository for multiple data sets and information on the species. Information housed in TAIR include the complete genome sequence along with gene structure, gene product information, gene expression, DNA and seed stocks, genome maps, genetic and physical markers, publications, and information about the Arabidopsis research community. There are several ecotypes of arabidopsis that have been useful in genetic research, and the natural variation has been used to identify loci important in both biotic and abiotic stress resistance.

Brachypodium Distachyon

Brachypodium distachyon is an experimental model grass that has many attributes that make it an excellent model for temperate cereals. Unlike wheat, a tetra or hexaploid species, brachypodium is diploid with a relatively small genome (~355 Mbp) with a short life-cycle, making genomic studies on it simpler.

Nicotiana Benthamiana

Nicotiana benthamiana is often considered a model organism for both plant-pathogen and transgenic studies. Because it is easily transformed with *Agrobacterium tumefaciens*, it is used to study both the expression of pathogen genes introduced into a plant or test new genetic cassette effects.

Genetically Modified Crops

Genetically modified (GM) foods are produced from organisms that have had changes introduced into their DNA using the methods of genetic engineering. Genetic engineering techniques allow for the introduction of new traits as well as greater control over traits than previous methods such as selective breeding and mutation breeding.

Genetically modifying plants is an important economic activity: in 2017, 89% of corn, 94% of soybeans, and 91% of cotton produced in the US were from genetically modified strains. Since the introduction of GM crops, yields have increased by 22%, and profits have increased to farmers, especially in the developing world, by 68%. An important side effect of GM crops has been decreased land requirements.

Commercial sale of genetically modified foods began in 1994, when Calgene first marketed its unsuccessful Flavr Savr delayed-ripening tomato. Most food modifications have primarily focused on cash crops in high demand by farmers such as soybean, corn, canola, and cotton. Genetically modified crops have been engineered for resistance to pathogens and herbicides and for better nutrient profiles. Other such crops include the economically important GM papaya which are resistant to the highly destructive Papaya ringspot virus, and the nutritionally improved golden rice (it is however still in development).

There is a scientific consensus that currently available food derived from GM crops poses no greater risk to human health than conventional food, but that each GM food needs to be tested on a case-by-case basis before introduction. Nonetheless, members of the public are much less likely than scientists to perceive GM foods as safe. The legal and regulatory status of GM foods varies by country, with some nations banning or restricting them, and others permitting them with widely differing degrees of regulation. There are still ongoing public concerns related to food safety, regulation, labeling, environmental impact, research methods, and the fact that some GM seeds are subject to intellectual property rights owned by corporations.

Modern Ways to Genetically Modify Plants

Genetic modification has been the cause for much research into modern plant genetics, and has also lead to the sequencing of many plant genomes. Today there are two predominant procedures of transforming genes in organisms: the "Gene gun" method and the *Agrobacterium* method.

Gene Gun Method

The gene gun method is also referred to as "biolistics" (ballistics using biological components). This technique is used for in vivo (within a living organism) transformation and has been especially useful in monocot species like corn and rice.This approach literally shoots genes into plant cells and plant cell chloroplasts. DNA is coated onto small particles of gold or tungsten approximately two micrometres in diameter. The particles are placed in a vacuum chamber and the plant tissue to be engineered is placed below the chamber. The particles are propelled at high velocity using a short pulse of high pressure helium gas, and hit a fine mesh baffle placed above the tissue while the DNA coating continues into any target cell or tissue.

Agrobacterium Method

Transformation via Agrobacterium has been successfully practiced in dicots, i.e. broad-leaf plants, such as soybeans and tomatoes, for many years. Recently it has been adapted and is now effective in monocots like grasses, including corn and rice. In general, the *Agrobacterium* method is considered preferable to the gene gun, because of a greater frequency of single-site insertions of the foreign DNA, which allows for easier monitoring. In this method, the tumor inducing (Ti) region is removed from the T-DNA (transfer DNA) and replaced with the desired gene and a marker, which is then inserted into the organism. This may involve direct inoculation of the tissue with a culture of transformed Agrobacterium, or inoculation following treatment with micro-projectile bombardment, which wounds the tissue. Wounding of the target tissue causes the release of phenolic compounds by the plant, which induces invasion of the tissue by Agrobacterium. Because of this, microprojectile bombardment often increases the efficiency of infection with Agrobacterium. The marker is used to find the organism which has successfully taken up the desired gene. Tissues of the organism are then transferred to a medium containing an antibiotic or herbicide, depending on which marker was used. The *Agrobacterium* present is also killed by the antibiotic. Only tissues expressing the marker will survive and possess the gene of interest. Thus, subsequent steps in the process will only use these surviving plants. In order to obtain whole plants from these tissues, they are grown under controlled environmental conditions in tissue culture.

This is a process of a series of media, each containing nutrients and hormones. Once the plants are grown and produce seed, the process of evaluating the progeny begins. This process entails selection of the seeds with the desired traits and then retesting and growing to make sure that the entire process has been completed successfully with the desired results.

BRYOLOGY

Bryology is the study of the plant order Bryophyta, which contains the sub classes of Mosses (Musci), Liverworts (Hepaticae) and Hornworts (Anthocerotophyta).

In taxonomic classification terms these simple plants are contained within the sub-Phylum of bryophytes and are the ancestors of the first land plants on earth. Bryophytes were the first land plants to emerge (during the Silurian period) around 460 million years ago from the single-celled aquatic eukaryiotic organisms contained within what was known as the 'Primordial Soup'.

The main differences between bryophytes and the more complicated plants that evolved from them (such as trees and flowers) is that:

- Bryophytes reproduce via spore dispersal instead of developing seeds.

- Bryophytes do not have any roots but instead have rudimentary root-like structures called rhizoids.

- Mosses, liverworts and hornworts are classified as nonvascular land plants meaning that they do not possess vascular tissue known as xylem. A xylem is made up of a cellular compound known as Lignin, and allows trees etc to uptake water via translocation. The lignin contained within a xylem results in the coarse tissue that makes trees woody; its absence in bryophytes means they are exclusively comprised of soft plant tissue.

Bryophyte Classification

Geographical Location and Habitat

There are thought to be around 15, 000 species of bryophytes, containing around 9,000 species of moss and around 6,000 species of liverworts and hornworts.

As Bryophytes themselves lack the necessary structures to uptake water, they are often found in damp conditions whereby they can absorb water through osmosis (Philips, 1980). As a result various mosses such as Bryum cryophyllum (found in the Northwest region of Canada) occur ubiquitously on rocks along rivers and lakes.

Mosses and Liverworts are found anywhere in the world where there is water, except in the sea itself. They can be found in streams, rivers, along lakeshores and even in deserts, and can even be sustained from inter-tidal sea spray; examples have even been found in Antarctica.

Reproduction in Bryophytes

During reproduction water is vital to bryophytes as it facilitates the dispersal of sper-matozoa's, or the reproductive cells of mosses and liverworts. For example, moss species reproduce via the dispersal of microscopic spores that land on a damp surface and then germinate. Mosses may forcibly eject spores through a small opening (known as a sphagnum); release them through four slits (Andraea), or as in the majority of moss species, through an opening when the lid of the fruit stalk.

Liverworts

Liverworts are either flat-lobed structures, which may superficially resemble succulent vascular species such as some cacti species, or have small leaves in rows of three. The means of reproduction is different to mosses in that capsules break open into four flaps, releasing the spores.

What you Need to be a Bryologist

Bryology is a specialist scientific study and requires certain equipment as well as an intimate knowledge and understanding of bryophyte anatomy, to identify the seemingly identical 9,000 species of moss and 6,000 species of liverworts.

Equipment

Although a good knowledge of bryophyte identification is needed to be a professional bryologist, basic equipment can allow anyone to enjoy the diverse world of bryology.

Some equipment you will need includes:

- A hand lens is essential in identifying the characteristic spore heads that distinguish moss species from each other. A X10 lens is useful for general studying but a X20 will be needed for fine detail.

- A identification guide is also crucial in studying bryophytes to compare the characteristics of the plant you have in front of you with the key in the book. The keys work by choosing bryophytes with similar structures and characteristics to the plant in front of you, ruling out incorrect species via a process of elimination until you arrive at a species that roughly resembles your specimen.

- A microscope will often be the only way of determining the species of some bryo-phytes, as the determining features such as spore heads, are virtually invisible to the naked eye. A high power compound microscope in the range of X40 to X400 will be needed to identify most small moss species.

What use is the Study of Bryology in the Real World?

Liverworts and lichens are biological indicators of clean air and only grow in abundance in areas with extremely pure air. For example the dense forests of the Pacific Northwest of America, particularly in Oregon are saturated with lichens and liverworts, and the air in the region is extremely clean. So in this way bryologists and other scientists can gain an immediate visual indication of the air quality in a given area. If bryophytes are absent from an area it may indicate poor air quality and therefore other environmental issues.

POMOLOGY

Pomology is a branch of botany that studies and cultivates fruit. The denomination fruticulture—introduced from Romance languages is also used.

Illustration of the 'Willermoz' pear by Alexandre Bivort
from Album de Pomologie.

Pomological research is mainly focused on the development, enhancement, cultivation and physiological studies of fruit trees. The goals of fruit tree improvement include enhancement of fruit quality, regulation of production periods, and reduction of production cost. One involved in the science of pomology is called a pomologist. Pomology has been an important area of research for centuries.

United States

During the mid-19th century in the United States, farmers were expanding fruit orchard programs in response to growing markets. At the same time, horticulturists from the USDA and agricultural colleges were bringing new varieties to the United States from foreign expeditions, and developing experimental lots for these fruits. In response to this increased interest and activity, USDA established the Division of Pomology in 1886 and named Henry E. Van Deman as chief pomologist. An important focus of the division was to publish illustrated accounts of new varieties and to disseminate research findings to fruit growers and breeders through special publications and annual reports. During this period Andrew Jackson Downing and his brother Charles were prominent in Pomology and Horticulture, producing *The Fruits and Fruit Trees of America*.

The introduction of new varieties required exact depiction of the fruit so that plant breeders could accurately document and disseminate their research results. Since the use of scientific photography was not widespread in the late 19th Century, USDA commissioned artists to create watercolor illustrations of newly introduced cultivars. Many of the watercolors were used for lithographic reproductions in USDA publications, such as the *Report of the Pomologist* and the *Yearbook of Agriculture*.

Today, the collection of approximately 7,700 watercolors is preserved in the National Agricultural Library's Special Collections, where it serves as a major historic and botanic resource to a variety of researchers, including horticulturists, historians, artists, and publishers. The study of pomology has somewhat dwindled over the past century.

References

- Paleobotany, entry: newworldencyclopedia.org, Retrieved 12 July, 2019

- Ethnobotany, entry: newworldencyclopedia.org, Retrieved 11 January, 2019

- Keddy, Paul A. (2007). Plants and Vegetation. Cambridge: Cambridge University Press. ISBN 978-0-521-86480-0

- What-is-bryology: actforlibraries.org, Retrieved 1 August, 2019

- Carroll & Salt. Ecology for Gardeners. Timber Press, Inc. Pp. Glossary, page 287. ISBN 0-88192-611-6

- Plant-anatomy, science, encyclopedia: factmonster.com, Retrieved 5 March, 2019

- Schulze, Ernst-Detlef; et al. (2005). Plant Ecology. Springer. Retrieved April 24, 2012. ISBN 3-540-20833-X

Plant Cell Biology

Plants cells are the eukaryotic cells which are present in green plants. There are different types of plant cells including parenchyma, collenchyma, sclerenchyma, etc. The chapter closely examines the key concepts related to plant cell biology such as these types of plant cells and the different organelles within them to provide an extensive understanding of the subject.

PLANT CELLS

Plant cells are eukaryotic cells that vary in several fundamental factors from other eukaryotic organisms. Both plant and animal cells contain nucleus along with similar organelles. One of the distinctive aspects of a plant cell is the presence of a cell wall outside the cell membrane.

Cell Structure and Function

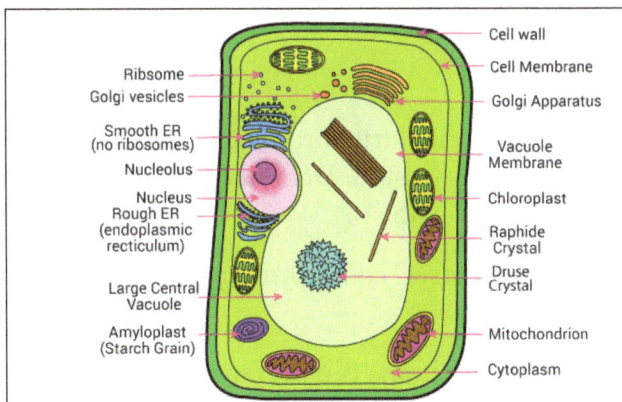

Diagram of Plant Cell

The plant cell is a rectangular and comparatively larger than the animal cells. Even though plant and animal cells are eukaryotic and share a few cell organelles, plant cells are quite distinct when compared to animal cell as they fulfil different functions. Some of these differences can be clearly understood when the cells are examined under an electron microscope.

Plant Cell Anatomy

Just like different organs within the body, plant cells have various components knows as cell organelles that perform a different function to sustain itself. These organelles include:

The Cell Wall

It is a rigid layer which is composed of cellulose, glycoproteins, lignin, pectin, and hemicellulose. It is located outside the cell membrane. It comprises proteins, polysaccharides, and cellulose. The primary function of the cell wall is to protect and provide structural support to the cell. The plant cell wall is also involved in protecting the cell against mechanical stress and to provide form and structure to the cell. It also filters the molecules passing into and outside the cell.

The formation of the cell wall is guided by microtubules. It consists of three layers, namely, primary, secondary and the middle lamella. The primary cell wall is formed by cellulose laid down by enzymes.

Cell Membrane

It is the semi-permeable membrane that is present within the cell wall. It is composed of a thin layer of protein and fat. The cell membrane plays an important role in regulating the entry and exit of specific substances within the cell. For instance, cell membrane keeps toxins from entering inside, while nutrients and essential minerals are transported across.

Nucleus

The nucleus is a membrane-bound structure that is present only in eukaryotic cells. The vital function of a nucleus is to store DNA or hereditary information required for cell division, metabolism, and growth.

1. Nucleolus: It manufactures cell's protein-producing structures and ribosomes.

2. Nucleopore: Nuclear membrane is perforated with holes called nucleopore that allows proteins and nucleic acids.

Plastids

They are membrane-bound organelles that have their own DNA. They are necessary to store starch, to carry out the process of photosynthesis. It is also used in the synthesis of many molecules which form the cellular building blocks. Some of the vital types of plastids and their functions are stated below.

Leucoplasts

They are found in non-photosynthetic tissues of plants. They are used for the storage of protein, lipid, and starch.

Chloroplasts

It is an elongated organelle enclosed by phospholipid membrane. The shape of the chloroplast is disk-shaped and the stroma is the fluid within the chloroplast that comprises a circular DNA. Each chloroplast contains a green coloured pigment called chlorophyll required for the process of photosynthesis. The chlorophyll absorbs light energy from the sun and uses it to transform carbon dioxide and water into glucose

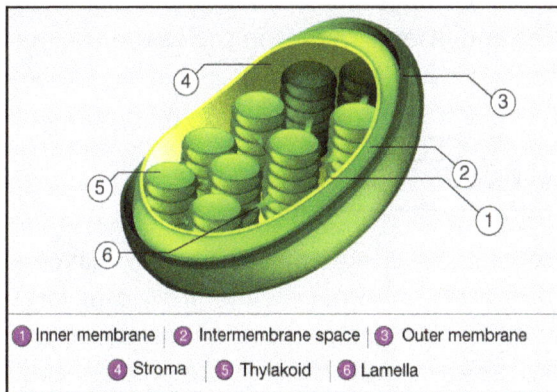

① Inner membrane | ② Intermembrane space | ③ Outer membrane
④ Stroma | ⑤ Thylakoid | ⑥ Lamella

Chromoplasts

They are heterogeneous, coloured plastids organelle which is responsible for pigment synthesis and for storage in photosynthetic eukaryotic organisms. Chromoplasts have red, orange and yellow coloured pigments which provide colour to all ripen fruits and flowers.

Central Vacuole

It occupies around thirty per cent of the cell's volume in a mature plant cell. Tonoplast is a membrane that surrounds central vacuole. The vital function of central vacuole apart from storage is to sustain turgid pressure against the cell wall. The central vacuole consists of cell sap. It is a mixture of salts, enzymes, and other substances.

Golgi Apparatus

They are found in all eukaryotic cells which are involved in distributing synthesized macromolecules to various parts of the cell.

Ribosomes

They are the smallest membrane-bound organelle which comprises RNA and protein.

They are the sites for protein synthesis, hence they are also referred to as the protein factories of the cell.

Mitochondria

They are the double-membraned organelles found in the cytoplasm of all eukaryotic cells. They provide energy by breaking down carbohydrate and sugar molecules, hence they are also referred to as the "Powerhouse of the cell."

Lysosome

Lysosomes are called as suicidal bags as they hold digestive enzymes in an enclosed membrane. They perform the function of cellular waste disposal by digesting worn-out organelles, food particles and foreign bodies in the cell.

TYPES OF PLANT CELLS

There is a wide range of cells found in plants. Within the leaves alone there is a variety of cells that perform different functions such as providing protection, photosynthesizing or transporting water. They same level of variety occurs in stems, roots and flowers. Here I describe only a few of the different types of cells from different parts of a plant.

Cells of Vascular Tissue

Vascular tissue in plants consists of the xylem and phloem. The xylem is responsible for transporting water through a plant. Xylem cells are dead, hollow tubes to allow water to move through plants as efficiently as possible.

The phloem is in charge of transporting sugars and other substances from leaves to roots. The cells of the phloem are also tubes but they are living cells. Phloem cells often have large pores to allow substances to move from one phloem cell to another.

Leaf Cells

Leaves are designed for photosynthesis. Many leaf cells are packed full of green organelles called chloroplasts. Chloroplasts are responsible for the production of sugars using the sun's energy and carbon dioxide.

Another structure that is extremely important for leaves and photosynthesis are stomata. Stomata are holes on the surface of leaves that can open and close to allow carbon dioxide into the leaf to be used in photosynthesis. Each stomata is made from two crescent-shaped cells that form a donut shape on the exterior of a leaf's skin.

Leaves also contain xylem and phloem cells which deliver water to leaves and take sugars away.

Cells of Stems and Roots

The majority of stem cells are a relatively featureless type of cell called a 'parenchyma' cell. Vascular tissues are bunched together into bundles with xylem cells and phloem cells sitting next to each other. Fiber cells, with thick cell walls, are not uncommon and are important for providing structural strength to plant stems.

Roots have many of the same cells as stems. Parenchyma cells are common and vascular cells once again form into bundles.

LEAF CELLS

A leaf cell is any cell found within a leaf. However, there are many different kinds of leaf cell, and each plays an integral role in the overall function of the leaf and the plant itself. A single leaf cell may be designed to simply photosynthesize, or create sugars from the energy in light. Other cells are designed to carry these sugars to the phloem, a specialized tube for transporting the sugars to the rest of the plant. Still other cells are specialized to carry water, which eventually form a rigid tube, the xylem. Another leaf cell is specifically designed to support the xylem and phloem into vascular bundles and transport substances to and from them.

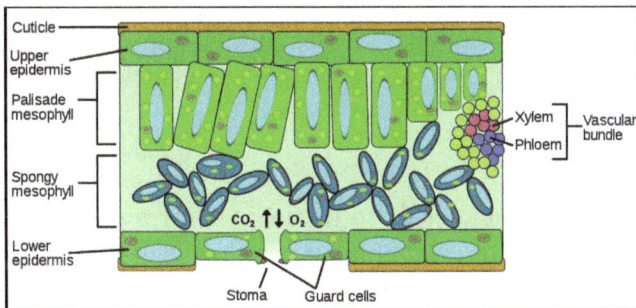

Leaf anatomy.

Types of Leaf Cell

Epidermis

An epidermal leaf cell is any cell which protects the outside of the leaf. These cells are often short and flattened, much like a square pancake. They form a protective layer over the leaf. They often produce waxy substances which protect the leaf from drying out or being attacked by insects. A leaf cell in the epidermis often lacks chloroplasts, the organelles responsible for creating sugar.

The upper and lower epidermis vary slightly. The upper epidermis, often exposed to direct sunlight, is often a thin layer of translucent cells. Below this are the cells responsible for photosynthesis, so they want to be as close to the light as possible while still being protected. The lower epidermis, on the other hand, is not responsible for protecting the plant from the harmful rays of sunlight. Instead, the lower epidermis has specialized cells for allowing air exchange. These small holes, called stoma, can be opened and closed by a specialized form of leaf cell.

Guard cells, as they are called, react to various condition inside and outside of the leaf, an open and close accordingly. It is through these stoma that the plant can exchange much needed carbon dioxide for the oxygen byproduct it is producing. Another important function of the stoma is transpiration. Through this process, water is passed out of the stoma and sucked up through the roots, bringing vital nutrients to the plant.

Palisade Mesophyll

The palisade mesophyll consists of a type of leaf cell specifically designed to carry out photosynthesis. These cells are absolutely packed with chlorophyll, and simply work their hardest to pump out as much sugar as they can. This sugar they release into the intracellular space, where it works its way to the next type of leaf cell.

Spongy Mesophyll

Spongy mesophyll is exactly what it sounds like: a loose matrix of structural mesophyll cells. These cells are not neatly packed into rows like the palisade cells. Rather, they form networks around bundles of vascular cells, and transport materials to and from the bundles. Like palisade mesophyll leaf cells, they can photosynthesize, but they carry additional functions as well. These two types of leaf cell give the leaf its green color.

Vascular Bundle

The last type of leaf cell is not specific to the leaf, as it travels the entire length of the plant. The cells around the xylem and phloem together make the vascular bundle. These highly specialized cells allow water and minerals to flow up from the roots, while transporting the products of photosynthesis to the entire plant. Like the arteries and veins of a human, they allow the organism to specialize functions in different parts of the body.

Guard Cell

Guard cells are specialized cells in the epidermis of leaves, stems and other organs that are used to control gas exchange. They are produced in pairs with a gap between them that forms a stomatal pore. The stomatal pores are largest when water is freely available and the guard cells turgid, and closed when water availability is critically low and the guard cells become flaccid. Photosynthesis depends on the diffusion of carbon dioxide (CO_2) from the air through the stomata into the mesophyll tissues. Oxygen (O_2), produced as a byproduct of photosynthesis, exits the plant via the stomata. When the stomata are open, water is lost by evaporation and must be replaced via the transpiration stream, with water taken up by the roots. Plants must balance the amount of CO_2 absorbed from the air with the water loss through the stomatal pores, and this is achieved by both active and passive control of guard cell turgor and stomatal pore size.

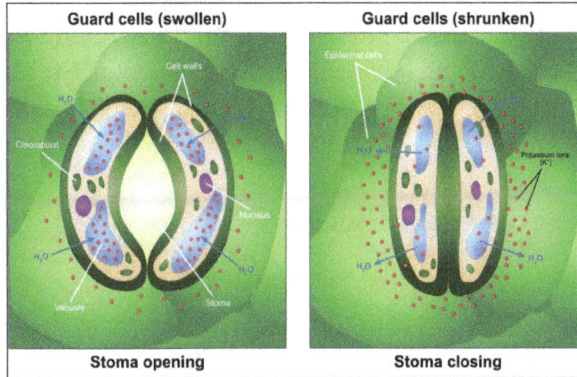

Opening and Closing of Stoma.

Guard Cell Function

Guard cells contain phototropins which are serine and threonine kinases mediated by light. Guard cells are cells surrounding each stoma. They help to regulate the rate of transpiration by opening and closing the stomata. They contain two light, oxygen, and voltage (LOV) domains, and are also part of the PAS domain superfamily. These phototropins trigger many responses such as phototropism, chloroplast movement, leaf expansion, and stomatal opening. Light is the main trigger for the opening or closing of stomata. Not much was previously known on how the photoreceptors work. To understand the mechanism to which phototropins work, an experiment on broad bean (*Vicia faba*) was done. Immunodetection and far-western blot analysis was used to determine blue light excites phototropin 1 and phototropin 2, which causes protein phosphate 1 to begin a phosphorylation cascade, which activates H^+-ATPase, a pump responsible for pumping H^+ ions out of the cell.

The phosphorylated H^+-ATPase allows the binding of a 14-3-3 protein, an autoinhibitory domain of H^+-ATPase, to the C terminus. Serine and threonine are then phosphorylated within the protein, which induces H^+-ATPase activity. The same experiment also found that upon phosphorylation, a 14-3-3 protein was bound to the phototropins before the H^+-ATPase had been phosphorylated. In a similar experiment they concluded that the binding of 14-3-3 protein to the phosphorylation site is essential to the activation of plasma membrane H^+-ATPase activity. This was done by adding phosphopeptides such as P-950, which inhibits the binding of 14-3-3 protein, to phosphorylated H^+-ATPase and observing the amino acid sequence. As protons are being pumped out, a negative electrical potential within the plasma membrane is formed. This hyperpolarization of the membrane allows for the accumulation of positively charged potassium

(K^+) ions and chloride (Cl^-) ions, which in turn, increases the solute concentration causing the water potential to decrease. The negative water potential allows for osmosis to occur in the guard cell, allowing the cell to become turgid. Opening and closure of the stomatal pore is mediated by changes in the turgor pressure of the two guard cells. The turgor pressure of guard cells is controlled by movements of large quantities of ions and sugars into and out of the guard cells. Guard cells have cell walls of varying thickness and differently oriented cellulose microfibers, causing them to bend outward when they are turgid, which in turn, causes stomata to open. Stomata close when there is an osmotic loss of water, occurring from the loss of K^+ to neighboring cells, mainly potassium (K^+) ions.

Water Loss and Water use Efficiency

Water stress (drought and salt stress) is one of the major environmental problems causing severe losses in agriculture and in nature. Drought tolerance of plants is mediated by several mechanisms that work together, including stabilizing and protecting the plant from damage caused by desiccation and also controlling how much water plants lose through the stomatal pores during drought. A plant hormone, abscisic acid (ABA), is produced in response to drought. A major type of ABA receptor has been identified. The plant hormone ABA causes the stomatal pores to close in response to drought, which reduces plant water loss via transpiration to the atmosphere and allows plants to avoid or slow down water loss during droughts. The use of drought tolerant crop plants would lead to a reduction in crop losses during droughts. Since guard cells control water loss of plants, the investigation on how stomatal opening and closure are regulated could lead to the development of plants with improved avoidance or slowing of desiccation and better water use efficiency. Research done Jean-Pierre Rona shows that ABA is the trigger for the closure of the stomatal opening.

To trigger this it activates the release of anions and potassium ions. This influx in anions causes a depolarization of the plasma membrane. This depolarization triggers potassium plus ions in the cell to leave the cell due to the unbalance in the membrane potential. This sudden change in ion concentrations cause the guard cell to shrink which causes the stomata to close which in turn decreases the amount of water lost. All this is a chain reaction according to her research. The increase in ABA causes there to be an increase in calcium ion concentration. Although at first they thought it was a coincidence they later discovered that this calcium increase is important. They found Ca^{2+} ions are involved in anion channel activation, which allows for anions to flow into the guard cell. They also are involved in prohibiting proton ATPase from correcting and stopping the membrane from being depolarized. To support their hypothesis that calcium was responsible for all these changes in the cell they did an experiment where they used proteins that inhibited the calcium ions for being produced. If their assumption that calcium is important in these processes they'd see that with the inhibitors they'd see less of the following things. Their assumption was correct and when the inhibitors were used they saw that the proton ATPase worked better to balance the depolarization. They also found that the flow of anions into the guard cells were not as strong. This is important for getting ions to flow into the guard cell. These two things are crucial in causing the stomatal opening to close preventing water loss for the plant.

Ion Uptake and Release

Ion uptake into guard cells causes stomatal opening: The opening of gas exchange pores requires the uptake of potassium ions into guard cells. Potassium channels and pumps have been identified and shown to function in the uptake of ions and opening of stomatal apertures. Ion release from guard cells causes stomatal pore closing: Other ion channels have been identified that mediate release of ions from guard cells, which results in osmotic water efflux from guard cells due to osmosis, shrinking of the guard cells, and closing of stomatal pores. Specialized potassium efflux channels participate in mediating release of potassium from guard cells. Anion channels were identified as important controllers of stomatal closing.

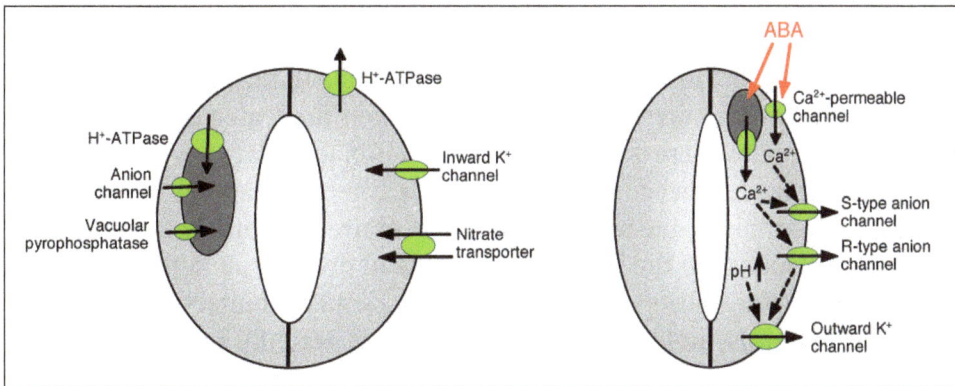

Ion channels and pumps regulating stomatal opening and closure.

Anion channels have several major functions in controlling stomatal closing: (a) They allow release of anions, such as chloride and malate from guard cells, which is needed for stomatal closing. (b) Anion channels are activated by signals that cause stomatal closing, for example by intracellular calcium and ABA. The resulting release of negatively charged anions from guard cells results in an electrical shift of the membrane to more positive voltages (depolarization) at the intracellular surface of the guard cell plasma membrane. This electrical depolarization of guard cells leads to activation of the outward potassium channels and the release of potassium through these channels. At least two major types of anion channels have been characterized in the plasma membrane: S-type anion channels and R-type anion channels.

Vacuolarion Transport

Vacuoles are large intracellular storage organelles in plants cells. In addition to the ion channels in the plasma membrane, vacuolar ion channels have important functions in regulation of stomatal opening and closure because vacuoles can occupy up to 90% of guard cell's volume. Therefore, a majority of ions are released from vacuoles when stomata are closed. Vascuolar K^+ (VK) channels and fast vacuolar channels can mediate K^+ release from vacuoles. Vacuolar K^+ (VK) channels are activated by elevation in the intracellular calcium concentration. Another type of calcium-activated channel, is the slow vacuolar (SV) channel. SV channels have been shown to function as cation channels that are permeable to Ca^{2+} ions, but their exact functions are not yet known in plants.

Guard cells control gas exchange and ion exchange through opening and closing. K^+ is one ion that flows both into and out of the cell, causing a positive charge to develop. Malate is one of the main anions used to counteract this positive charge, and it is moved through the AtALMT6 ion channel. AtALMT6 is an aluminum-activated malate transporter that is found in guard cells, specifically in the vacuoles. This transport channel was found to cause either an influx or efflux of malate depending on the concentrations of calcium. In a study by Meyer et al, patch clamp experiments were conducted on mesophyll vacuoles from arabidopsis rdr6-11 (WT) and arabidopsis that were overexpressing AtALMT6-GFP. It was found from these experiments that in the WT there were only small currents when calcium ions were introduced, while in the AtALMT6-GFP mutant a huge inward rectifying current was observed. When the transporter is knocked out from guard cell vacuoles there is a significant reduction in malate flow current. The current goes from a huge inward current to not much different than the WT, and Meyer et al hypothesized that this is due to residual malate concentrations in the vacuole. There is also a similar response in the knockout mutants to drought as in the WT. There was no phenotypic difference observed between the knockout mutants, the wild type, or the AtALMT6-GFP mutants, and the exact cause for this is not fully known.

Signal Transduction

Guard cells perceive and process environmental and endogenous stimuli such as light, humidity, CO_2 concentration, temperature, drought, and plant hormones to trigger cellular responses resulting in stomatal opening or closure. These signal transduction pathways determine for example how quickly a plant will lose water during a drought period. Guard cells have become a model for single cell signaling. Using *Arabidopsis thaliana*, the investigation of signal processing in single guard cells has become open to the power of genetics. Cytosolic and nuclear proteins and chemical messengers that function in stomatal movements have been identified that mediate the transduction of environmental signals thus controlling CO_2 intake into plants and plant water loss. Research on guard cell signal transduction mechanisms is producing an understanding of how plants can improve their response to drought stress by reducing plant water loss. Guard cells also provide an excellent model for basic studies on how a cell integrates numerous kinds of input signals to produce a response (stomatal opening or closing). These responses require coordination of numerous cell biological processes in guard cells, including signal reception, ion channel and pump regulation, membrane trafficking, transcription, cytoskeletal rearrangements and more. A challenge for future research is to assign the functions of some of the identified proteins to these diverse cell biological processes.

Development

During the development of plant leaves, the specialized guard cells differentiate from "guard mother cells". The density of the stomatal pores in leaves is regulated by environmental signals, including increasing atmospheric CO_2 concentration, which reduces the density of stomatal pores in the surface of leaves in many plant species by presently unknown mechanisms. The genetics of stomatal development can be directly studied by imaging of the leaf epidermis using a microscope. Several major control proteins that function in a pathway mediating the development of guard cells and the stomatal pores have been identified.

PLANT CELL ORGANELLES

A typical plant cell is made up of cytoplasm and organelles. In fact, all the organelles (except nucleus) and subcellular structures are present in the cytoplasm, which is enclosed by protective layers (the cell wall and cell membrane). Scientific studies have been done regarding the cell organelles and their functions. Each of the organelles of a plant cell has specific function, without which the cell cannot operate properly.

The plant cell is protected from the surrounding environment by the cell wall and cell membrane. Note that these two are surface structures and not cell organelles. They not only give shape, support and strength to the cell, but also aid in transportation. When it comes to the organelles found in a plant cell, they are more or less similar to animal cells, except that the latter lacks chloroplasts, that are responsible for photosynthesis. Following is a list of organelles found in a plant cell.

Nucleus

Nucleus (plural nuclei) is a highly specialized cell organelle, which stores the genetic component (chromosomes) of the particular cell. It serves as the main administrative center of the cell by coordinating the metabolic processes like cell growth, cell division and protein synthesis. Together, the nucleus along with its contents is referred to as nucleoplasm.

Plastids (Chloroplasts)

Plastids is a collective term for organelles that carry pigments. In a plant cell, chloroplasts are the most prominent forms of plastids that contain the green chlorophyll pigment. Because of these chloroplast plastids, a plant cell has the ability to undergo photosynthesis in the presence of sunlight, water and carbon dioxide to synthesize its own food.

Ribosomes

Ribosomes are plant organelles that comprise of proteins (40 percent) and ribonucleic acid or RNA (60 percent). They are responsible for the synthesis of proteins. Inside the cell, a ribosome may occur freely (free ribosome) or it may be attached to another organelle, endoplasmic reticulum (bound ribosome). Each ribosome consists of two parts, a larger sub-unit and a smaller sub-unit.

Mitochondria

Mitochondria (singular mitochondrion) are large, spherical or rod-shaped organelles present in the cytoplasm of the plant cell. They break down the complex carbohydrates and sugars into usable forms for the plant. A mitochondrion contains certain enzymes that are essential for supply of energy to the plant cell. Hence, these cell organelles are also known as the powerhouse of the cell.

Golgi Body

A golgi body is also referred to as golgi complex or golgi apparatus. It plays a major role in transporting chemical substances in and out of the cell. After the endoplasmic reticulum synthesizes lipids and proteins, golgi body alters and prepares them for exporting outside the cell. Arranged in a sac-like pattern, this organelle is located near the cell nucleus.

Endoplasmic Reticulum

Endoplasmic reticulum (ER) is the connecting link between the nucleus and cytoplasm of the plant cell. Basically, it is a network of interconnected, convoluted sacs present in the cytoplasm. Based on the presence or absence of ribosomes, ER can be of smooth or rough types. The former type lacks ribosomes, while the latter is covered with ribosomes. Overall, endoplasmic reticulum serves to manufacture, store and transport, structure for glycogen, proteins, steroids, and other compounds.

Vacuoles

Vacuoles are the membrane-bound, storage organelles that help in regulating turgor pressure of the plant cell. In a plant cell, there can be more than one vacuole. However, the centrally located vacuole is larger than the others, which stores all sorts of chemical compounds. Vacuoles also assist in intracellular digestion of complex molecules and excretion of waste products.

Peroxisomes

Peroxisomes are cytoplasmic organelles of the plant cell, which contain certain oxidative enzymes. These enzymes are used for the metabolic breakdown of fatty acids into simple sugar forms. Another important function of peroxisomes is to help chloroplasts in undergoing the photorespiration process.

Cytoskeleton

The cytoskeleton occupies a large volume in a cell. It consists of filaments and tubules which extend throughout the cytoplasm. Its main function is to give shape and support within the cell. It helps to keep the cell organelles in place.

Well. This was a brief information regarding plant cell organelles, their structure and their functions. Other parts of the plant cell include microfilaments (structural component) and plasmodesmata (connecting tubes between the cells). As we have seen above, the coordination of cell organelles is crucial for carrying out the physiological and biochemical functionalities of the plant.

References

- Plant-cell, biology: byjus.com, Retrieved 23 February, 2019

- Shimazaki, Ken-ichiro; Doi, Michio; Assmann, Sarah M.; Kinoshita, Toshinori (2007). "Light Regulation of Stomatal Movement". Annual Review of Plant Biology. 58 (1): 219–247. Doi:10.1146/annurev.arplant.57.032905.105434. PMID 17209798

- Plant-cells, cells, micro: basicbiology.net, Retrieved 22 May, 2019

- Leaf-cell: biologydictionary.net, Retrieved 13 May, 2019

- Plant-cell-organelles: biologywise.com, Retrieved 2 March, 2019

Plant Tissues

The combination of similar cells and their extracellular matrix which carries out a specific function is known as a tissue. The different types of plant tissues include vascular tissues, epidermal tissues and ground tissues. The chapter closely examines these types of plant tissues to provide an extensive understanding of the subject.

Plants are immobile and henceforth have been provided with tissues made up of dead cells, which provide structural strength. They have to endure unfavorable environmental situations like strong winds, storms, floods etc.

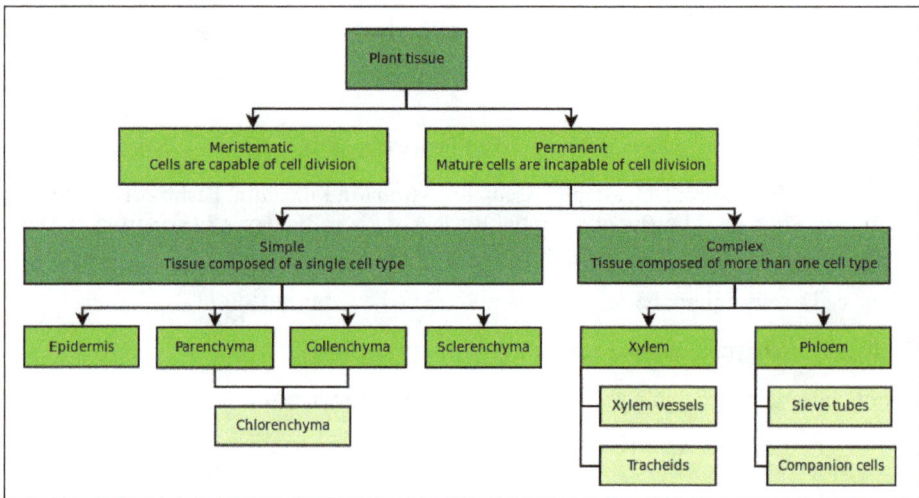

Plant Tissue System.

A tissue is a cluster of cells, that are alike in configuration and work together to attain a specific function. Different types of plant tissues include permanent and meristematic tissues.

Meristematic Tissue

These tissues have the capability to develop by swift division. They assist in the major growth of the vegetation. Growth in length and growth in diameter of the plant are carried about by these cells. The Meristematic cells are cubical, living cells with a big nucleus. These cells are meticulously crammed with no intercellular spaces. Depending on the section where the meristematic tissues are existing, they are categorized as meristems intercalary, lateral and apical.

- Apical meristem is existent at the growing tips or apical of stems and roots. Apical meristem upsurges the length of the plant.

- Lateral meristem is existent in the radial portion of the stem or root. Lateral meristem upsurges the thickness of the plant.

- Intercalary meristem is found at the internodes or at the base of the leaves. Intercalary meristem upsurges the size of the internode.

Old meristematic cells lose the capability to distribute and convert into permanent tissues. This procedure of capturing up a permanent function, size, and shape is termed as differentiation.

Permanent Tissues

Those cells which have lost their ability to distribute but are specialised to offer elasticity, flexibility and strength to the plant. These tissues can be additionally categorised into:

- Simple Permanent Tissue: They can be classified into sclerenchyma, collenchyma and parenchyma based on their purpose.

- Complex Permanent Tissue: These tissues include of phloem and xylem. Xylem is valuable for the transportation of water and solvable constituents. Xylem is made up of xylem parenchyma, fibres,vessels and tracheids. Phloem is valuable in the transportation of food particles. Phloem consists of phloem parenchyma, phloem fibres, companion cells, sieve cells and sieve tubes.

Parenchyma

These are alive, polygonal cells with a big central vacuole, and have intercellular spaces amidst them. Parenchymatous cells create the ground tissue and pith.

- Parenchyma comprising of chloroplasts are termed as chlorenchyma. The chlorenchyma helps in photosynthesis.

- Parenchyma which comprise of big air voids are called aerenchyma. Buoyancy is the main purpose the aerenchyma serves.

- Some parenchymatous cells perform as storage chambers for starch in vegetable and fruits.

Collenchyma

These are stretched out living cells with minute intercellular gaps. Their cell walls are made up of pectin and cellulose. Collenchyma is found in the marginal regions of leaves and stems and offers flexibility with the structural framework and mechanical support in plants.

Sclerenchyma

These are elongated, dead cells with a lignin deposits in their cell wall. They have no intercellular gaps. Sclerenchyma is found in the covering of seeds and nuts, around the vascular tissues in stems and the veins of leaves. Sclerenchyma provides strength to the plant.

Xylem

It helps in the transport of dissolved substances and water all through the plant. The diverse components of the xylem include vessels, tracheids, xylem fibres and xylem parenchyma. Xylem fibres and Tracheids are made up of lignin, which provides structural support to the plant.

Phloem

This tissue helps in the transportation of food all through the plant. The diverse elements of phloem include phloem fibres, sieve tubes, phloem parenchyma and companion cells.

Protective Tissues

These provide fortification to the plant. They include the cork and epidermis:

- Epidermis: Its a layer of cell that makes up making up an outer casing of all the structures in the plant. The stomata perforates the epidermis at certain places. The stomata help in loss of water and gaseous exchange.

- Cork: This is the external protective tissue which substitutes the epidermal cells in mature stems and roots. Cork cells are lifeless and lack intercellular gaps. Their cell walls are coagulated by suberin which makes them impervious to gas and Water Molecules.

VASCULAR TISSUE

Vascular tissue is a complex conducting tissue, formed of more than one cell type, found in vascular plants. The primary components of vascular tissue are the xylem and phloem. These two tissues transport fluid and nutrients internally. There are also two meristems associated with vascular tissue: the vascular cambium and the cork cambium. All the vascular tissues within a particular plant together constitute the vascular tissue system of that plant.

Cross section of celery stalk, showing vascular bundles, which include both phloem and xylem.

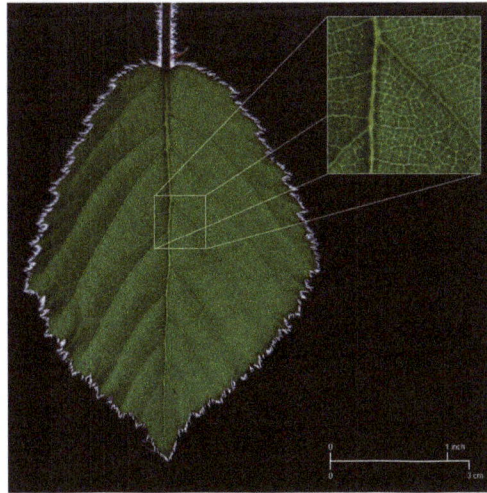

Detail of the vasculature of a bramble leaf.

The cells in vascular tissue are typically long and slender. Since the xylem and phloem function in the conduction of water, minerals, and nutrients throughout the plant, it is not surprising that their form should be similar to pipes. The individual cells of phloem are connected end-to-end, just as the sections of a pipe might be. As the plant grows, new vascular tissue differentiates in the growing tips of the plant. The new tissue is aligned with existing vascular tissue, maintaining its connection throughout the plant. The vascular tissue in plants is arranged in long, discrete strands called vascular bundles. These bundles include both xylem and phloem, as well as supporting and protective cells. In stems and roots, the xylem typically lies closer to the interior of the stem with phloem towards the exterior of the stem. In the stems of some Asterales dicots, there may be phloem located inwardly from the xylem as well.

Between the xylem and phloem is a meristem called the vascular cambium. This tissue divides off cells that will become additional xylem and phloem. This growth increases the girth of the plant, rather than its length. As long as the vascular cambium continues to produce new cells, the plant will continue to grow more stout. In trees and other plants that develop wood, the vascular cambium allows the expansion of vascular tissue that produces woody growth. Because this growth ruptures the epidermis of the stem, woody plants also have a cork cambium that develops among the phloem. The cork cambium gives rise to thickened cork cells to protect the surface of the plant and reduce water loss. Both the production of wood and the production of cork are forms of secondary growth.

In leaves, the vascular bundles are located among the spongy mesophyll. The xylem is oriented toward the adaxial surface of the leaf (usually the upper side), and phloem is oriented toward the abaxial surface of the leaf. This is why aphids are typically found on the undersides of the leaves rather than on the top, since the phloem transports sugars manufactured by the plant and they are closer to the lower surface.

Xylem

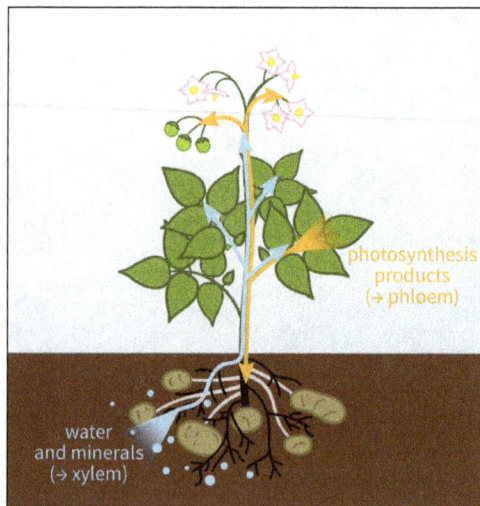

Xylem (blue) transports water and minerals from the roots upwards.

Xylem is one of the two types of transport tissue in vascular plants, phloem being the other. The basic function of xylem is to transport water from roots to stems and leaves, but it also transports nutrients. The word "xylem" is derived from the Greek word ξύλον (*xylon*), meaning "wood"; the best-known xylem tissue is wood, though it is found throughout a plant. The term was introduced by Carl Nägeli in 1858.

Xylem appeared early in the history of terrestrial plant life. Fossil plants with anatomically preserved xylem are known from the Silurian (more than 400 million years ago), and trace fossils resembling individual xylem cells may be found in earlier Ordovician rocks.The earliest true and recognizable xylem consists of tracheids with a helical-annular reinforcing layer added to the cell wall. This is the only type of xylem found in the earliest vascular plants, and this type of cell continues to be found in the protoxylem (first-formed xylem) of all living groups of vascular plants. Several groups of plants later developed pitted tracheid cells independently through convergent evolution. In living plants, pitted tracheids do not appear in development until the maturation of the metaxylem (following the protoxylem).

In most plants, pitted tracheids function as the primary transport cells. The other type of vascular element, found in angiosperms, is the vessel element. Vessel elements are joined end to end to form vessels in which water flows unimpeded, as in a pipe. The presence of xylem vessels is considered to be one of the key innovations that led to the success of the angiosperms. However, the occurrence of vessel elements is not restricted to angiosperms, and they are absent in some archaic or "basal" lineages of the angiosperms: (e.g., Amborellaceae, Tetracentraceae, Trochodendraceae, and Winteraceae), and their secondary xylem is described by Arthur Cronquist as "primitively vesselless". Cronquist considered the vessels of Gnetum to be convergent with those of angiosperms. Whether the absence of vessels in basal angiosperms is a primitive condition is contested, the alternative hypothesis states that vessel elements originated in a precursor to the angiosperms and were subsequently lost.

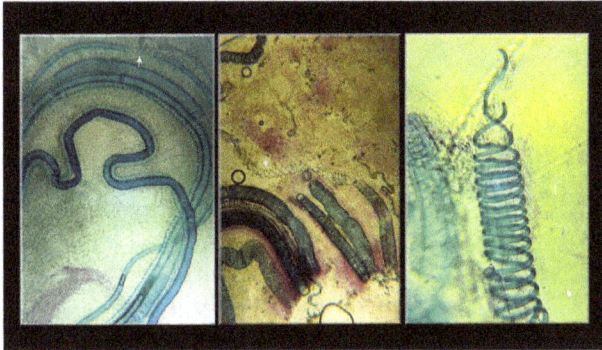

Photos showing xylem elements in the shoot of a fig tree (Ficus alba): crushed in hydrochloric acid, between slides and cover slips.

The high CO_2 levels of Silurian-Devonian times, when plants were first colonizing land, meant that the need for water was relatively low. As CO_2 was withdrawn from the atmosphere by plants, more water was lost in its capture, and more elegant transport mechanisms evolved. As water transport mechanisms, and waterproof cuticles, evolved, plants could survive without being continually covered by a film of water. This transition from poikilohydry to homoiohydry opened up new potential for colonization. Plants then needed a robust internal structure that held long narrow channels for transporting water from the soil to all the different parts of the above-soil plant, especially to the parts where photosynthesis occurred.

During the Silurian, CO_2 was readily available, so little water needed expending to acquire it. By the end of the Carboniferous, when CO_2 levels had lowered to something approaching today's, around 17 times more water was lost per unit of CO_2 uptake. However, even in these "easy" early days, water was at a premium, and had to be transported to parts of the plant from the wet soil to avoid desiccation. This early water transport took advantage of the *cohesion-tension* mechanism inherent in water.

Water has a tendency to diffuse to areas that are drier, and this process is accelerated when water can be wicked along a fabric with small spaces. In small passages, such as that between the plant cell walls (or in tracheids), a column of water behaves like rubber – when molecules evaporate from one end, they pull the molecules behind them along the channels. Therefore, transpiration alone provided the driving force for water transport in early plants. However, without dedicated transport vessels, the cohesion-tension mechanism cannot transport water more than about 2 cm, severely limiting the size of the earliest plants. This process demands a steady supply of water from one end, to maintain the chains; to avoid exhausting it, plants developed a waterproof cuticle. Early cuticle may not have had pores but did not cover the entire plant surface, so that gas exchange could continue. However, dehydration at times was inevitable; early plants cope with this by having a lot of water stored between their cell walls, and when it comes to it sticking out the tough times by putting life "on hold" until more water is supplied.

A banded tube from the late Silurian/early Devonian. The bands are difficult to see on this specimen, as an opaque carbonaceous coating conceals much of the tube. Bands are just visible in places on the left half of the image.

To be free from the constraints of small size and constant moisture that the parenchymatic transport system inflicted, plants needed a more efficient water transport system. During the early Silurian, they developed specialized cells, which were lignified (or bore similar chemical compounds) to avoid implosion; this process coincided with cell death, allowing their innards to be emptied and water to be passed through them. These wider, dead, empty cells were a million times more conductive than the inter-cell method, giving the potential for transport over longer distances, and higher CO_2 diffusion rates.

The earliest macrofossils to bear water-transport tubes are Silurian plants placed in the genus *Cooksonia*. The early Devonian pretracheophytes *Aglaophyton* and *Horneophyton* have structures very similar to the hydroids of modern mosses. Plants continued to innovate new ways of reducing the resistance to flow within their cells, thereby increasing the efficiency of their water transport. Bands on the walls of tubes, in fact apparent from the early Silurian onwards, are an early improvisation to aid the easy flow of water. Banded tubes, as well as tubes with pitted ornamentation on their walls, were lignified and, when they form single celled conduits, are considered to be *tracheids*. These, the "next generation" of transport cell design, have a more rigid structure than hydroids, allowing them to cope with higher levels of water pressure. Tracheids may have a single evolutionary origin, possibly within the hornworts, uniting all tracheophytes (but they may have evolved more than once).

Water transport requires regulation, and dynamic control is provided by stomata. By adjusting the amount of gas exchange, they can restrict the amount of water lost through transpiration. This is an important role where water supply is not constant, and indeed stomata appear to have evolved before tracheids, being present in the non-vascular hornworts.

An endodermis probably evolved during the Silu-Devonian, but the first fossil evidence for such a structure is Carboniferous. This structure in the roots covers the water transport tissue and regulates ion exchange (and prevents unwanted pathogens etc. from entering the water transport system). The endodermis can also provide an upwards pressure, forcing water out of the roots when transpiration is not enough of a driver.

Once plants had evolved this level of controlled water transport, they were truly homoiohydric, able to extract water from their environment through root-like organs rather than relying on a film of surface moisture, enabling them to grow to much greater size. As a result of their independence from their surroundings, they lost their ability to survive desiccation – a costly trait to retain.

During the Devonian, maximum xylem diameter increased with time, with the minimum diameter remaining pretty constant. By the middle Devonian, the tracheid diameter of some plant lineages (Zosterophyllophytes) had plateaued. Wider tracheids allow water to be transported faster, but the overall transport rate depends also on the overall cross-sectional area of the xylem bundle itself. The increase in vascular bundle thickness further seems to correlate with the width of plant axes, and plant height; it is also closely related to the appearance of leaves and increased stomatal density, both of which would increase the demand for water.

While wider tracheids with robust walls make it possible to achieve higher water transport pressures, this increases the problem of cavitation. Cavitation occurs when a bubble of air forms within a vessel, breaking the bonds between chains of water molecules and preventing them from pulling more water up with their cohesive tension. A tracheid, once cavitated, cannot have its embolism removed and return to service (except in a few advanced angiosperms which have developed a mechanism of doing so). Therefore, it is well worth plants' while to avoid cavitation occurring. For this reason, pits in tracheid walls have very small diameters, to prevent air entering and allowing bubbles to nucleate. Freeze-thaw cycles are a major cause of cavitation. Damage to a tracheid's wall almost inevitably leads to air leaking in and cavitation, hence the importance of many tracheids working in parallel.

Cavitation is hard to avoid, but once it has occurred plants have a range of mechanisms to contain the damage. Small pits link adjacent conduits to allow fluid to flow between them, but not air – although ironically these pits, which prevent the spread of embolisms, are also a major cause of them. These pitted surfaces further reduce the flow of water through the xylem by as much as 30%. Conifers, by the Jurassic, developed an ingenious improvement, using valve-like structures to isolate cavitated elements. These torus-margo structures have a blob floating in the middle of a donut; when one side depressurizes the blob is sucked into the torus and blocks further flow. Other plants simply accept cavitation; for instance, oaks grow a ring of wide vessels at the start of each spring, none of which survive the winter frosts. Maples use root pressure each spring to force sap upwards from the roots, squeezing out any air bubbles.

Growing to height also employed another trait of tracheids – the support offered by their lignified walls. Defunct tracheids were retained to form a strong, woody stem, produced in most instances by a secondary xylem. However, in early plants, tracheids were too mechanically vulnerable, and retained a central position, with a layer of tough sclerenchyma on the outer rim of the stems. Even when tracheids do take a structural role, they are supported by sclerenchymatic tissue.

Tracheids end with walls, which impose a great deal of resistance on flow; vessel members have perforated end walls, and are arranged in series to operate as if they were one continuous vessel. The function of end walls, which were the default state in the Devonian, was probably to avoid embolisms. An embolism is where an air bubble is created in a tracheid. This may happen as a result of freezing, or by gases dissolving out of solution. Once an embolism is formed, it usually cannot be removed; the affected cell cannot pull water up, and is rendered useless.

End walls excluded, the tracheids of prevascular plants were able to operate under the same hydraulic conductivity as those of the first vascular plant, *Cooksonia*.

The size of tracheids is limited as they comprise a single cell; this limits their length, which in turn limits their maximum useful diameter to 80 μm. Conductivity grows with the fourth power of diameter, so increased diameter has huge rewards; *vessel elements*, consisting of a number of cells, joined at their ends, overcame this limit and allowed larger tubes to form, reaching diameters of up to 500 μm, and lengths of up to 10 m.

Vessels first evolved during the dry, low CO_2 periods of the late Permian, in the horsetails, ferns and Selaginellales independently, and later appeared in the mid Cretaceous in angiosperms and gnetophytes. Vessels allow the same cross-sectional area of wood to transport around a hundred times more water than tracheids! This allowed plants to fill more of their stems with structural fibers, and also opened a new niche to vines, which could transport water without being as thick as the tree they grew on. Despite these advantages, tracheid-based wood is a lot lighter, thus cheaper to make, as vessels need to be much more reinforced to avoid cavitation.

Development

Xylem development can be described by four terms: *centrarch, exarch, endarch* and *mesarch*. As it develops in young plants, its nature changes from *protoxylem* to *metaxylem* (i.e. from *first xylem* to *after xylem*). The patterns in which protoxylem and metaxylem are arranged is important in the study of plant morphology.

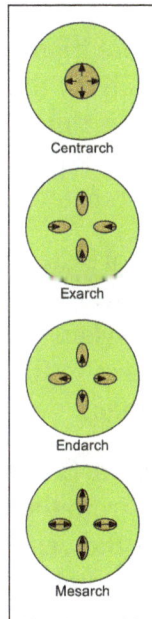

Patterns of xylem development: xylem in brown; arrows show direction of development from protoxylem to metaxylem.

Protoxylem and Metaxylem

As a young vascular plant grows, one or more strands of primary xylem form in its stems and roots. The first xylem to develop is called 'protoxylem'. In appearance protoxylem is usually distinguished by narrower vessels formed of smaller cells. Some of these cells have walls which contain thickenings in the form of rings or helices. Functionally, protoxylem can extend: the cells are able to grow in size and develop while a stem or root is elongating. Later, 'metaxylem' develops in the strands of xylem. Metaxylem vessels and cells are usually larger; the cells have thickenings which are typically either in the form of ladderlike transverse bars (scalariform) or continuous sheets except for holes or pits (pitted). Functionally, metaxylem completes its development after elongation ceases when the cells no longer need to grow in size.

Patterns of Protoxylem and Metaxylem

There are four main patterns to the arrangement of protoxylem and metaxylem in stems and roots:

- *Centrarch* refers to the case in which the primary xylem forms a single cylinder in the center of the stem and develops from the center outwards. The protoxylem is thus found in the central core and the metaxylem in a cylinder around it. This pattern was common in early land plants, such as "rhyniophytes", but is not present in any living plants.

The other three terms are used where there is more than one strand of primary xylem:

- *Exarch* is used when there is more than one strand of primary xylem in a stem or root, and the xylem develops from the outside inwards towards the center, i.e. centripetally. The metaxylem is thus closest to the center of the stem or root and the protoxylem closest to the periphery. The roots of vascular plants are normally considered to have exarch development.

- *Endarch* is used when there is more than one strand of primary xylem in a stem or root, and the xylem develops from the inside outwards towards the periphery, i.e. centrifugally. The protoxylem is thus closest to the center of the stem or root and the metaxylem closest to the periphery. The stems of seed plants typically have endarch development.

- *Mesarch* is used when there is more than one strand of primary xylem in a stem or root, and the xylem develops from the middle of a strand in both directions. The metaxylem is thus on both the peripheral and central sides of the strand with the protoxylem between the metaxylem (possibly surrounded by it). The leaves and stems of many ferns have mesarch development.

Structure

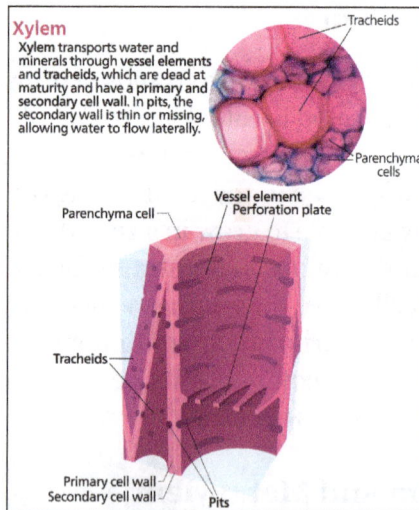

Xylem
Xylem transports water and minerals through vessel elements and tracheids, which are dead at maturity and have a primary and secondary cell wall. In pits, the secondary wall is thin or missing, allowing water to flow laterally.

Tracheids

Parenchyma cells

Vessel element
Perforation plate

Parenchyma cell

Tracheids

Primary cell wall
Secondary cell wall   Pits

Cross section of some xylem cells.

The most distinctive xylem cells are the long tracheary elements that transport water. Tracheids and vessel elements are distinguished by their shape; vessel elements are shorter, and are connected together into long tubes that are called *vessels*.

Xylem also contains two other cell types: parenchyma and fibers.

Xylem can be found:

- In vascular bundles, present in non-woody plants and non-woody parts of woody plants.

- In secondary xylem, laid down by a meristem called the vascular cambium in woody plants.

- As part of a stelar arrangement not divided into bundles, as in many ferns.

In transitional stages of plants with secondary growth, the first two categories are not mutually exclusive, although usually a vascular bundle will contain *primary xylem* only. The branching pattern exhibited by xylem follows Murray's law.

Primary and Secondary Xylem

Primary xylem is formed during primary growth from procambium. It includes protoxylem and metaxylem. Metaxylem develops after the protoxylem but before secondary xylem. Metaxylem has wider vessels and tracheids than protoxylem.

Secondary xylem is formed during secondary growth from vascular cambium. Although secondary xylem is also found in members of the gymnosperm groups Gnetophyta and Ginkgophyta and to a lesser extent in members of the Cycadophyta, the two main groups in which secondary xylem can be found are:

- Conifers (*Coniferae*): there are approximately 600 known species of conifers. All species have secondary xylem, which is relatively uniform in structure throughout this group. Many conifers become tall trees: the secondary xylem of such trees is used and marketed as softwood.

- Angiosperms (*Angiospermae*): there are approximately 250,000 known species of angiosperms. Within this group secondary xylem is rare in the monocots. Many non-monocot angiosperms become trees, and the secondary xylem of these is used and marketed as hardwood.

Main Function – Upwards Water Transport

The xylem, vessels and tracheids of the roots, stems and leaves are interconnected to form a continuous system of water-conducting channels reaching all parts of the plants. The system transports water and soluble mineral nutrients from the roots throughout the plant. It is also used to replace water lost during transpiration and photosynthesis. Xylem sap consists mainly of water and inorganic ions, although it can also contain a number of organic chemicals as well. The transport is passive, not powered by energy spent by the tracheary elements themselves, which are dead by maturity and no longer have living contents. Transporting sap upwards becomes more difficult as the height of a plant increases and upwards transport of water by xylem is considered to limit the maximum height of trees. Three phenomena cause xylem sap to flow:

- Pressure flow hypothesis: Sugars produced in the leaves and other green tissues are kept in the phloem system, creating a solute pressure differential versus the xylem system carrying a far lower load of solutes- water and minerals. The phloem pressure can rise to several MPa, far higher than atmospheric pressure. Selective inter-connection between these systems allows this high solute concentration in the phloem to draw xylem fluid upwards by negative pressure.

- Transpirational pull: Similarly, the evaporation of water from the surfaces of mesophyll cells to the atmosphere also creates a negative pressure at the top of a plant. This causes millions of minute menisci to form in the mesophyll cell wall. The resulting surface tension causes a negative pressure or tension in the xylem that pulls the water from the roots and soil.

- Root pressure: If the water potential of the root cells is more negative than that of the soil, usually due to high concentrations of solute, water can move by osmosis into the root from the soil. This causes a positive pressure that forces sap up the xylem towards the leaves. In some circumstances, the sap will be forced from the leaf through a hydathode in a phenomenon known as guttation. Root pressure is highest in the morning before the stomata open and allow transpiration to begin. Different plant species can have different root pressures even in a similar environment; examples include up to 145 kPa in *Vitis riparia* but around zero in *Celastrus orbiculatus*.

The primary force that creates the capillary action movement of water upwards in plants is the adhesion between the water and the surface of the xylem conduits. Capillary action provides the force that establishes an equilibrium configuration, balancing gravity. When transpiration removes water at the top, the flow is needed to return to the equilibrium.

Transpirational pull results from the evaporation of water from the surfaces of cells in the leaves. This evaporation causes the surface of the water to recess into the pores of the cell wall. By capillary action, the water forms concave menisci inside the pores. The high surface tension of water pulls the concavity outwards, generating enough force to lift water as high as a hundred meters from ground level to a tree's highest branches.

Transpirational pull requires that the vessels transporting the water be very small in diameter; otherwise, cavitation would break the water column. And as water evaporates from leaves, more is drawn up through the plant to replace it. When the water pressure within the xylem reaches extreme levels due to low water input from the roots (if, for example, the soil is dry), then the gases come out of solution and form a bubble – an embolism forms, which will spread quickly to other adjacent cells, unless bordered pits are present (these have a plug-like structure called a torus, that seals off the opening between adjacent cells and stops the embolism from spreading).

Cohesion-tension Theory

The *cohesion-tension theory* is a theory of intermolecular attraction that explains the process of water flow upwards (against the force of gravity) through the xylem of plants. It was proposed in 1894 by John Joly and Henry Horatio Dixon. Despite numerous objections, this is the most widely accepted theory for the transport of water through a plant's vascular system based on the classical research of Dixon-Joly, Eugen Askenasy and Dixon.

Water is a polar molecule. When two water molecules approach one another, the slightly negatively charged oxygen atom of one forms a hydrogen bond with a slightly positively charged hydrogen atom in the other. This attractive force, along with other intermolecular forces, is one of the principal factors responsible for the occurrence of surface tension in liquid water. It also allows plants to draw water from the root through the xylem to the leaf.

Water is constantly lost through transpiration from the leaf. When one water molecule is lost another is pulled along by the processes of cohesion and tension. Transpiration pull, utilizing capillary action and the inherent surface tension of water, is the primary mechanism of water movement in plants. However, it is not the only mechanism involved. Any use of water in leaves forces water to move into them.

Transpiration in leaves creates tension (differential pressure) in the cell walls of mesophyll cells. Because of this tension, water is being pulled up from the roots into the leaves, helped by cohesion (the pull between individual water molecules, due to hydrogen bonds) and adhesion (the stickiness between water molecules and the hydrophilic cell walls of plants). This mechanism of water flow works because of water potential (water flows from high to low potential), and the rules of simple diffusion.

Over the past century, there has been a great deal of research regarding the mechanism of xylem sap transport; today, most plant scientists continue to agree that the *cohesion-tension theory* best explains this process, but multiforce theories that hypothesize several alternative mechanisms have been suggested, including longitudinal cellular and xylem osmotic pressure gradients, axial potential gradients in the vessels, and gel- and gas-bubble-supported interfacial gradients.

Measurement of Pressure

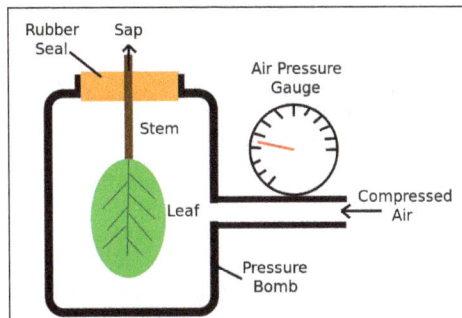

A diagram showing the setup of a pressure bomb.

Until recently, the differential pressure (suction) of transpirational pull could only be measured indirectly, by applying external pressure with a pressure bomb to counteract it. When the technology to perform direct measurements with a pressure probe was developed, there was initially some doubt about whether the classic theory was correct, because some workers were unable to demonstrate negative pressures. More recent measurements do tend to validate the classic theory, for the most part. Xylem transport is driven by a combination of transpirational pull from above and root pressure from below, which makes the interpretation of measurements more complicated.

Phloem

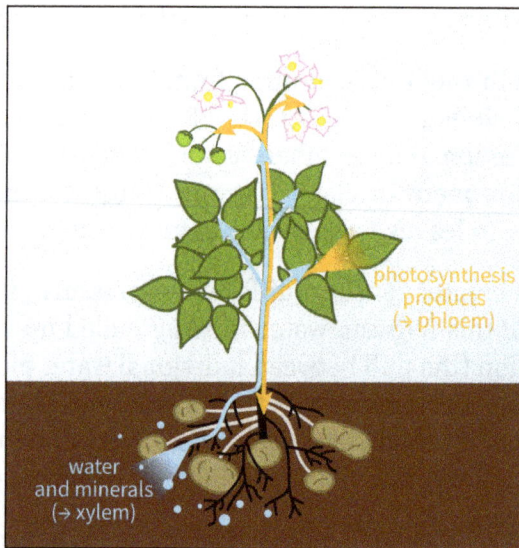

Phloem (orange) transports products of photosynthesis to various parts of the plant.

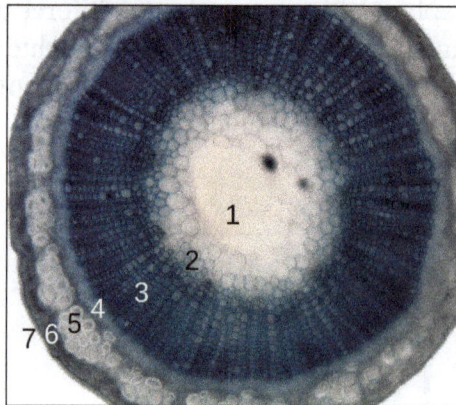

Cross-section of a flax plant stem:
Pith Protoxylem Metaxylem I Phloem I Sclerenchyma (bast fibre) Cortex Epidermis.

In vascular plants, phloem is the living tissue that transports the soluble organic compounds made during photosynthesis and known as *photosynthates*, in particular the sugar sucrose, to parts of the plant where needed. This transport process is called translocation. In trees, the phloem is the innermost layer of the bark,. The term was introduced by Nägeli in 1858.

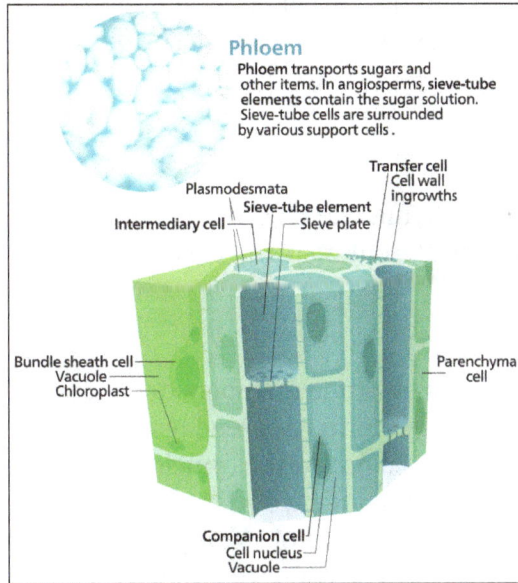

Cross section of some phloem cells.

Phloem tissue consists of conducting cells, generally called sieve elements, parenchyma cells, including both specialized companion cells or albuminous cells and unspecialized cells and supportive cells, such as fibres and sclereids.

Conducting Cells (Sieve Elements)

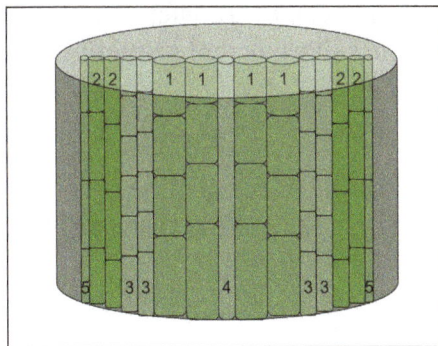

Simplified phloem and companion cells
Xylem Phloem Cambium Pith Companion Cells.

Sieve elements are the type of cell that are responsible for transporting sugars throughout the plant. At maturity they lack a nucleus and have very few organelles, so they rely on companion cells or albuminous cells for most of their metabolic needs. Sieve tube cells do contain vacuoles and other organelles, such as ribosomes, before they mature, but these generally migrate to the cell wall and dissolve at maturity; this ensures there is little to impede the movement of fluids. One of the few organelles they do contain at maturity is the rough endoplasmic reticulum, which can be found at the plasma membrane, often nearby the plasmodesmata that connect them to their companion or albuminous cells. All sieve cells have groups of pores at their ends that grow from modified and enlarged plasmodesmata, called sieve areas. The pores are reinforced by platelets of a polysaccharide called callose.

Parenchyma Cells

They are of two types, aerenchyma and chlorenchyma. Other parenchyma cells within the phloem are generally undifferentiated and used for food storage.

Companion Cells

The metabolic functioning of sieve-tube members depends on a close association with the *companion cells*, a specialized form of parenchyma cell. All of the cellular functions of a sieve-tube element are carried out by the (much smaller) companion cell, a typical nucleate plant cell except the companion cell usually has a larger number of ribosomes and mitochondria. The dense cytoplasm of a companion cell is connected to the sieve-tube element by plasmodesmata. The common sidewall shared by a sieve tube element and a companion cell has large numbers of plasmodesmata.

There are two types of companion cells:

- Ordinary companion cells, which have smooth walls and few or no plasmodesmatal connections to cells other than the sieve tube.

- Transfer cells, which have much-folded walls that are adjacent to non-sieve cells, allowing for larger areas of transfer. They are specialized in scavenging solutes from those in the cell walls that are actively pumped requiring energy.

Albuminous Cells

Albuminous cells have a similar role to companion cells, but are associated with sieve cells only and are hence found only in seedless vascular plants and gymnosperms.

Supportive Cells

Although its primary function is transport of sugars, phloem may also contain cells that have a mechanical support function. These generally fall into two categories: fibres and sclereids. Both cell types have a secondary cell wall and are therefore dead at maturity. The secondary cell wall increases their rigidity and tensile strength.

Fibres

Bast fibres are the long, narrow supportive cells that provide tension strength without limiting flexibility. They are also found in xylem, and are the main component of many textiles such as paper, linen, and cotton.

Sclereids

Sclereids are irregularly shaped cells that add compression strength but may reduce flexibility to some extent. They also serve as anti-herbivory structures, as their irregular shape and hardness will increase wear on teeth as the herbivores chews. For example, they are responsible for the gritty texture in pears, and in winter bears.

Function

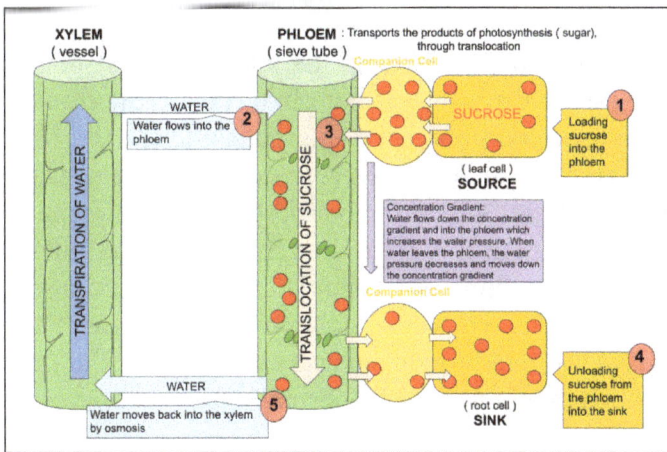

Unlike xylem (which is composed primarily of dead cells), the phloem is composed of still-living cells that transport sap. The sap is a water-based solution, but rich in sugars made by photosynthesis. These sugars are transported to non-photosynthetic parts of the plant, such as the roots, or into storage structures, such as tubers or bulbs.

During the plant's growth period, usually during the spring, storage organs such as the roots are sugar sources, and the plant's many growing areas are sugar sinks. The movement in phloem is multidirectional, whereas, in xylem cells, it is unidirectional (upward).

After the growth period, when the meristems are dormant, the leaves are sources, and storage organs are sinks. Developing seed-bearing organs (such as fruit) are always sinks. Because of this multi-directional flow, coupled with the fact that sap cannot move with ease between adjacent sieve-tubes, it is not unusual for sap in adjacent sieve-tubes to be flowing in opposite directions.

While movement of water and minerals through the xylem is driven by negative pressures (tension) most of the time, movement through the phloem is driven by positive hydrostatic pressures. This process is termed *translocation*, and is accomplished by a process called phloem loading and *unloading*.

Phloem sap is also thought to play a role in sending informational signals throughout vascular plants. "Loading and unloading patterns are largely determined by the conductivity and number of plasmodesmata and the position-dependent function of solute-specific, plasma membrane transport proteins. Recent evidence indicates that mobile proteins and RNA are part of the plant's long-distance communication signaling system. Evidence also exists for the directed transport and sorting of macromolecules as they pass through plasmodesmata."

Organic molecules such as sugars, amino acids, certain hormones, and even messenger RNAs are transported in the phloem through sieve tube elements.

Girdling

Because phloem tubes are located outside the xylem in most plants, a tree or other plant can be killed by stripping away the bark in a ring on the trunk or stem. With the phloem destroyed, nutrients cannot reach the roots, and the tree/plant will die. Trees located in areas with animals such as beavers are vulnerable since beavers chew off the bark at a fairly precise height. This process is known as girdling, and can be used for agricultural purposes. For example, enormous fruits and vegetables seen at fairs and carnivals are produced via girdling. A farmer would place a girdle at the base of a large branch, and remove all but one fruit/vegetable from that branch. Thus, all the sugars manufactured by leaves on that branch have no sinks to go to but the one fruit/vegetable, which thus expands to many times its normal size.

When the plant is an embryo, vascular tissue emerges from procambium tissue, which is at the center of the embryo. Protophloem itself appears in the mid-vein extending into the cotyledonary node, which constitutes the first appearance of a leaf in angiosperms, where it forms continuous strands. The hormone auxin, transported by the protein PIN1 is responsible for the growth of those protophloem strands, signaling the final identity of those tissues. SHORTROOT(SHR), and microRNA165/166 also participate in that process, while Callose Synthase 3(CALS3), inhibits the locations where SHORTROOT(SHR), and microRNA165 can go.

In the embryo, root phloem develops independently in the upper hypocotyl, which lies between the embryonic root, and the cotyledon.

In an adult, the phloem originates, and grows outwards from, meristematic cells in the vascular cambium. Phloem is produced in phases. *Primary* phloem is laid down by the apical meristem and develops from the procambium. *Secondary* phloem is laid down by the vascular cambium to the inside of the established layer(s) of phloem.

In some eudicot families (Apocynaceae, Convolvulaceae, Cucurbitaceae, Solanaceae, Myrtaceae, Asteraceae, Thymelaeaceae), phloem also develops on the inner side of the vascular cambium; in this case, a distinction between external phloem and internal phloem or intraxylary phloem is made. Internal phloem is mostly primary, and begins differentiation later than the external phloem and protoxylem, though it is not without exceptions. In some other families (Amaranthaceae, Nyctaginaceae, Salvadoraceae), the cambium also periodically forms inward strands or layers of phloem, embedded in the xylem: Such phloem strands are called included phloem or interxylary phloem.

EPIDERMAL TISSUE

Epidermal tissue system is the outermost covering of plants. It consists of epidermis, stomata and epidermal outgrowths. Epidermis is generally composed of single layer of parenchymatous cells compactly arranged without intercellular spaces. But it is interrupted by stomata. In leaves some specialized cells which surround the stomata are called the guard cells.

Epidermal Tissue System

Epidermal tissue system is the outermost covering of plants. It consists of epidermis, stomata and epidermal outgrowths. Epidermis is generally composed of single layer of parenchymatous cells compactly arranged without intercellular spaces. But it is interrupted by stomata. In leaves some specialized cells which surround the stomata are called the guard cells. Chloroplasts are present only in the guard cells of the epidermis. Other epidermal cells usually do not have chloroplasts. The outer wall of epidermis is usually covered by cuticle.

Stoma is a minute pore surrounded by two guard cells. The stomata occur mainly in the epidermis of leaves. In some plants such as sugarcane, the guard cells are bounded by some special cells. They are distinct from other epidermal cells. These cells are called subsidiary or accessory cells. Trichomes and root hairs are some epidermal outgrowths. The unicellular or multicellular appendages that originate from the epidermal cells are called trichomes. Trichomes may be branched or unbranched. Rhizodermis has two types of epidermal cells - long cells and short cells. The short cells are called trichoblasts. Root hairs are produced from these trichoblasts.

Functions of Epidermal Tissue System

- This tissue system in the shoot checks excessive loss of water due to the presence of cuticle.

- Epidermis protects the underlying tissues.

- Stomata involve in transpiration and gaseous exchange.

- Trichomes are also helpful in the dispersal of seeds and fruits.

- Root hairs absorb water and mineral salts from the soil.

GROUND TISSUE

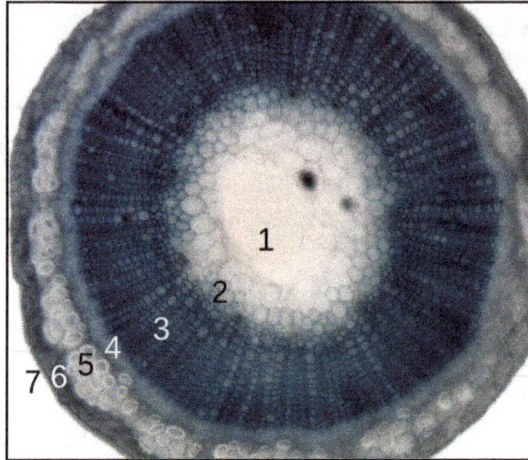

The ground tissue of plants includes all tissues that are neither dermal nor vascular. It can be divided into three types based on the nature of the cell walls.

1. Parenchyma cells have thin primary walls and usually remain alive after they become mature. Parenchyma forms the "filler" tissue in the soft parts of plants, and is usually present in cortex, pericycle, pith, and medullary rays in primary stem and root.

2. Collenchyma cells have thin primary walls with some areas of secondary thickening. Collenchyma provides extra mechanical and structural support, particularly in regions of new growth.

3. Sclerenchyma cells have thick lignified secondary walls and often die when mature. Sclerenchyma provides the main structural support to a plant.

Parenchyma is a versatile ground tissue that generally constitutes the "filler" tissue in soft parts of plants. It forms, among other things, the cortex and pith of stems, the cortex of roots, the mesophyll of leaves, the pulp of fruits, and the endosperm of seeds. Parenchyma cells are living cells and may remain meristematic at maturity—meaning that they are capable of cell division if stimulated. They have thin and flexible cellulose cell walls, and are generally polyhedral when close-packed, but can be roughly spherical when isolated from their neighbours. They have large central vacuoles, which allow the cells to store and regulate ions, waste products, and water. Tissue specialised for food storage is commonly formed of parenchyma cells.

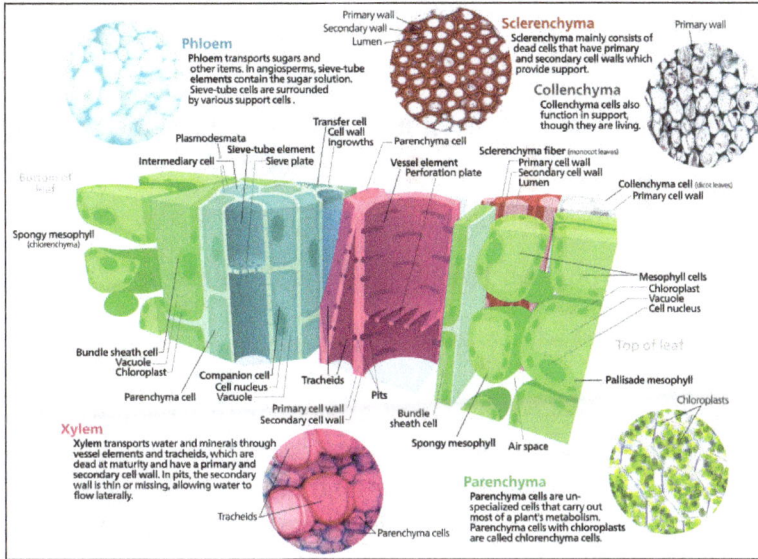

Phloem
Phloem transports sugars and other items. In angiosperms, sieve-tube elements contain the sugar solution. Sieve-tube cells are surrounded by various support cells.

Primary wall
Secondary wall
Lumen

Sclerenchyma
Sclerenchyma mainly consists of dead cells that have primary and secondary cell walls which provide support.

Primary wall

Collenchyma
Collenchyma cells also function in support, though they are living.

Transfer cell
Cell wall ingrowths
Plasmodesmata
Intermediary cell
Sieve-tube element
Sieve plate

Parenchyma cell
Vessel element
Perforation plate

Sclerenchyma fiber (monocot leaves)
Primary cell wall
Secondary cell wall
Lumen

Collenchyma cell (dicot leaves)
Primary cell wall

Spongy mesophyll
(chlorenchyma)

Mesophyll cells
Chloroplast
Vacuole
Cell nucleus

Bundle sheath cell
Vacuole
Chloroplast
Parenchyma cell

Companion cell
Cell nucleus
Vacuole

Tracheids

Pits

Primary cell wall
Secondary cell wall

Bundle sheath cell
Spongy mesophyll

Top of leaf

Palisade mesophyll
Chloroplasts

Xylem
Xylem transports water and minerals through vessel elements and tracheids, which are dead at maturity and have a primary and secondary cell wall. In pits, the secondary wall is thin or missing, allowing water to flow laterally.

Tracheids

Air space

Parenchyma cells

Parenchyma
Parenchyma cells are un-specialized cells that carry out most of a plant's metabolism. Parenchyma cells with chloroplasts are called chlorenchyma cells.

Cross section of a leaf showing various ground tissue types.

Parenchyma cells have a variety of functions:

- Their main function is to repair.

- In leaves, they form the mesophyll and are responsible for photosynthesis and the exchange of gases. Parenchyma cells in the mesophyll of leaves are specialised parenchyma cells called chlorenchyma cells (parenchyma cells with chloroplasts).

- Storage of starch, protein, fats, oils and water in roots, tubers (e.g. potatoes), seed endosperm (e.g. cereals) and cotyledons (e.g. pulses and peanuts).

- Secretion (e.g. the parenchyma cells lining the inside of resin ducts).

- Wound repair and the potential for renewed meristematic activity.

- Other specialised functions such as aeration (aerenchyma) provides buoyancy and helps aquatic plants float.

- Chlorenchyma cells carry out photosynthesis and manufacture food.

The shape of parenchyma cells varies with their function. In the spongy mesophyll of a leaf, parenchyma cells range from near-spherical and loosely arranged with large intercellular spaces, to branched or stellate, mutually interconnected with their neighbours at the ends of their arms to form a three-dimensional network, like in the red kidney bean *Phaseolus vulgaris* and other mesophytes. These cells, along with the epidermal guard cells of the stoma, form a system of air spaces and chambers that regulate the exchange of gases. In some works, the cells of the leaf epidermis are regarded as specialised parenchymal cells, but the modern preference has long been to classify the epidermis as plant dermal tissue, and parenchyma as ground tissue.

Shapes of parenchyma:

- Polyhedral.

- Stellate (found in stem of plants and have well developed air spaces between them).

- Elongated (found in pallisade tissue of leaf).

- Lobed (found in spongy and pallisade mesophyyll tissue of some plants).

Collenchyma

Collenchyma tissue is composed of elongated cells with irregularly thickened walls. They provide structural support, particularly in growing shoots and leaves. Collenchyma tissue makes up things such as the resilient strands in stalks of celery. Collenchyma cells are usually living, and have only a thick primary cell wall made up of cellulose and pectin. Cell wall thickness is strongly affected by mechanical stress upon the plant. The walls of collenchyma in shaken plants (to mimic the effects of wind etc.), may be 40–100% thicker than those not shaken.

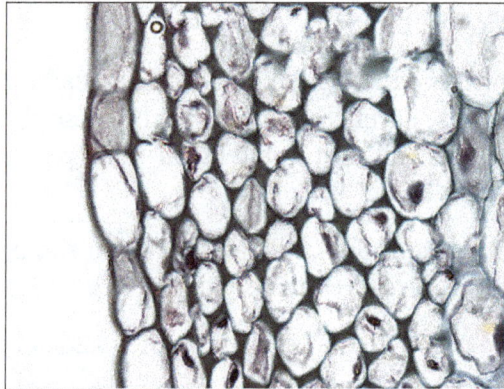
Cross section of collenchyma cells.

There are four main types of collenchyma:

- Angular collenchyma (thickened at intercellular contact points).

- Tangential collenchyma (cells arranged into ordered rows and thickened at the tangential face of the cell wall).

- Annular collenchyma (uniformly thickened cell walls).

- Lacunar collenchyma (collenchyma with intercellular spaces).

Collenchyma cells are most often found adjacent to outer growing tissues such as the vascular cambium and are known for increasing structural support and integrity.

The first use of "collenchyma" was by Link who used it to describe the sticky substance on *Bletia* (Orchidaceae) pollen. Complaining about Link's excessive nomenclature, Schleiden stated mockingly that the term "collenchyma" could have more easily been used to describe elongated sub-epidermal cells with unevenly thickened cell walls.

Sclerenchyma

Sclerenchyma is the tissue which makes the plant hard and stiff. Sclerenchyma is the supporting tissue in plants. Two types of sclerenchyma cells exist: fibers and sclereids. Their cell walls consist of cellulose, hemicellulose and lignin. Sclerenchyma cells are the principal supporting cells in plant tissues that have ceased elongation. Sclerenchyma fibers are of great economic importance, since they constitute the source material for many fabrics (e.g. [flax]) hemp, jute, and ramie).

Unlike the collenchyma, mature sclerenchyma is composed of dead cells with extremely thick cell walls (secondary walls) that make up to 90% of the whole cell volume. It is the hard, thick walls that make sclerenchyma cells important strengthening and supporting elements in plant parts that have ceased elongation. The difference between fibers and sclereids is not always clear: transitions do exist, sometimes even within the same plant.

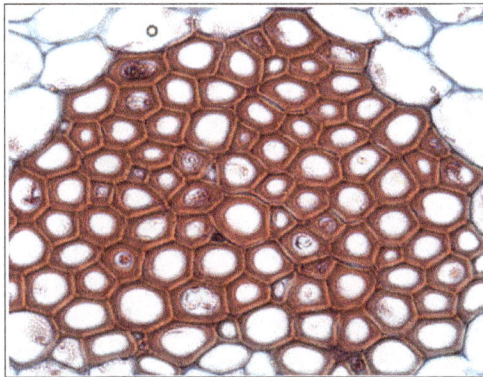

Cross section of sclerenchyma fibers.

Fibers or bast are generally long, slender, so-called prosenchymatous cells, usually occurring in strands or bundles. Such bundles or the totality of a stem's bundles are colloquially called fibers. Their high load-bearing capacity and the ease with which they can be processed has since antiquity made them the source material for a number of things, like ropes, fabrics and mattresses. The fibers of flax (*Linum usitatissimum*) have been known in Europe and Egypt for more than 3,000 years, those of hemp (*Cannabis sativa*) in China for just as long. These fibers, and those of jute (*Corchorus capsularis*) and ramie (*Boehmeria nivea*, a nettle), are extremely soft and elastic and are especially well suited for the processing to textiles. Their principal cell wall material is cellulose.

Contrasting are hard fibers that are mostly found in monocots. Typical examples are the fiber of many grasses, agaves (sisal: *Agave sisalana*), lilies (*Yucca* or *Phormium tenax*), *Musa textilis* and others. Their cell walls contain, besides cellulose, a high proportion of lignin. The load-bearing capacity of *Phormium tenax* is as high as 20–25 kg/mm², the same as that of good steel wire (25 kg/ mm²), but the fibre tears as soon as too great a strain is placed upon it, while the wire distorts and does not tear before a strain of 80 kg/mm². The thickening of a cell wall has been studied in *Linum*. Starting at the centre of the fiber, the thickening layers of the secondary wall are deposited one after the other. Growth at both tips of the cell leads to simultaneous elongation. During development the layers of secondary material seem like tubes, of which the outer one is always longer and older than the next. After completion of growth, the missing parts are supplemented, so that the wall is evenly thickened up to the tips of the fibers.

Fibers usually originate from meristematic tissues. Cambium and procambium are their main centers of production. They are usually associated with the xylem and phloem of the vascular bundles. The fibers of the xylem are always lignified, while those of the phloem are cellulosic. Reliable evidence for the fibre cells' evolutionary origin from tracheids exists. During evolution the strength of the tracheid cell walls was enhanced, the ability to conduct water was lost and the size of the pits was reduced. Fibers that do not belong to the xylem are bast (outside the ring of cambium) and such fibers that are arranged in characteristic patterns at different sites of the shoot. The term "sclerenchyma" (originally *Sclerenchyma*) was introduced by Mettenius in 1865.

Parenchyma

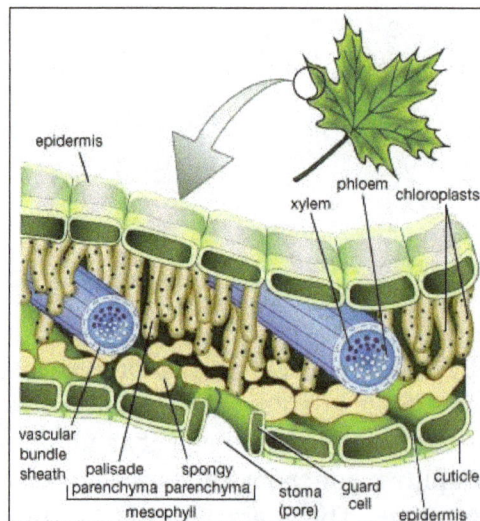

Parenchyma is composed of living cells that are thin-walled, unspecialized in structure, and therefore adaptable, with differentiation, to various functions. The cells are found in many places throughout plant bodies and, given that they are alive, are actively involved in photosynthesis, secretion, food storage, and other activities of plant life. Parenchyma is one of the three main types of ground, or fundamental, tissue in plants, together with sclerenchyma (dead support tissues with thick walls) and collenchyma (living support tissues with irregular walls).

Parenchyma makes up the chloroplast-laden mesophyll (internal layers) of leaves and the cortex (outer layers) and pith (innermost layers) of stems and roots; it also forms the soft tissues of fruits. Cells of this type are also contained in xylem and phloem as transfer cells and as the bundle sheaths that surround the vascular strands. Parenchyma tissue may be compact or have extensive spaces between the cells.

Leaf mesophyll composed of parenchyma tissue. The elongated palisade parenchyma contains the largest number of chloroplasts per cell and is the primary site of photosynthesis in many plants. The irregular spongy parenchyma also contains chloroplasts and facilitates the passage of gases through its many intercellular spaces.

Collenchyma

In plants, Collenchyma tissue of living elongated cells with irregular cell walls. Collenchyma cells have thick deposits of cellulose in their cell walls and appear polygonal in cross section. The strength of the tissue results from these thickened cell walls and the longitudinal interlocking of the cells. Collenchyma may form cylinders or occur as discrete strands and is one of the three ground, or fundamental, tissues in plants, together with parenchyma (living thin-walled tissue) and sclerenchyma (dead support tissue with thick cell walls).

An important feature of collenchyma is that it is extremely plastic—the cells can extend and thus adjust to increased growth of the organ. The tissue is found chiefly in the cortex of stems and in leaves and is the primary supporting tissue for many herbaceous plants. In plants with secondary growth, the collenchyma tissue is only temporarily functional and becomes crushed as woody tissue develops. It often constitutes the ridges and angles of stems and commonly borders the veins in eudicot leaves. The "strings" in stalks of celery are a notable example of collenchyma tissue.

Sclerenchyma

In plants, Sclerenchyma tissue composed of any of various kinds of hard woody cells. Mature sclerenchyma cells are usually dead cells that have heavily thickened secondary walls containing lignin. The cells are rigid and nonstretchable and are usually found in nongrowing regions of plant bodies, such as the bark or mature stems. Sclerenchyma is one of the three types of ground, or fundamental, tissue in plants; the other two types are parenchyma (living thin-walled tissue) and collenchyma (living support tissue with irregular walls). Sclerenchyma cells occur in many different shapes and sizes, but two main types occur: fibres and sclereids.

The three types of ground, or fundamental, tissue in plants. Parenchyma tissue is composed of thin-walled cells and makes up the photosynthetic tissue in leaves, the pulp of fruits, and the endosperm of many seeds. Collenchyma cells mainly form supporting tissue and have irregular cell walls. They are found mainly in the cortex of stems and in leaves. The major function of sclerenchyma is support. Unlike collenchyma, mature cells of this tissue are generally dead and have thick walls containing lignin. Their size, shape, and structure vary greatly.

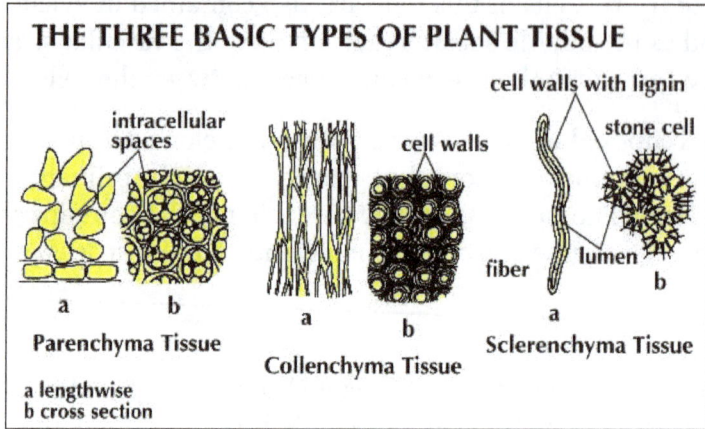

THE THREE BASIC TYPES OF PLANT TISSUE

Parenchyma Tissue

Collenchyma Tissue

Sclerenchyma Tissue

a lengthwise
b cross section

Ground tissue.

Fibres are greatly elongated cells whose long, tapering ends interlock, thus providing maximum support to a plant. They often occur in bundles or strands and can be found almost anywhere in the plant body, including the stem, the roots, and the vascular bundles in leaves. Many of these fibres, including seed hairs, leaf fibres, and bast fibres, are important sources of raw material for textiles and other woven goods.

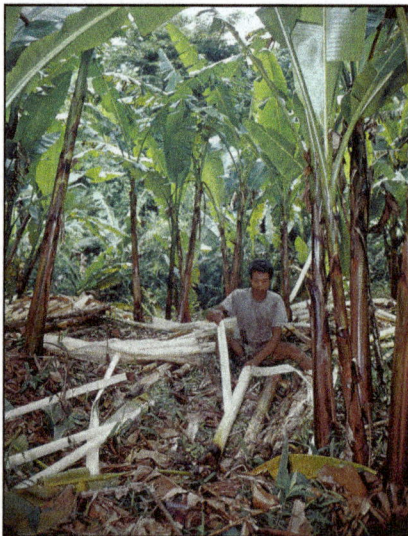

A worker stripping fibre from abaca (Musa textilis)
in the Philippines.

Sclereids are extremely variable in shape and are present in various tissues of the plant, such as the periderm, cortex, pith, xylem, and phloem. They also occur in leaves and fruits and constitute the hard shell of nuts and the outer hard coat of many seeds. Sometimes known as stone cells, sclereids are also responsible for the gritty texture of pears and guavas.

MERISTEM

A meristem is the tissue in most plants containing undifferentiated cells (meristematic cells), found in zones of the plant where growth can take place. Meristematic cells give rise to various organs of a plant and are responsible for growth.

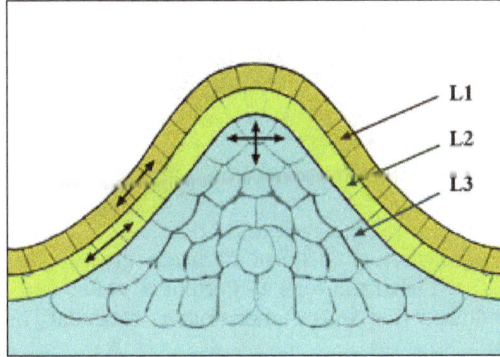

Tunica-Corpus model of the apical meristem (growing tip). The epidermal (L1) and subepidermal (L2) layers form the outer layers called the tunica. The inner L3 layer is called the corpus. Cells in the L1 and L2 layers divide in a sideways fashion, which keeps these layers distinct, whereas the L3 layer divides in a more random fashion.

Differentiated plant cells generally cannot divide or produce cells of a different type. Meristematic cells are incompletely or not at all differentiated, and are capable of continued cellular division. Therefore, cell division in the meristem is required to provide new cells for expansion and differentiation of tissues and initiation of new organs, providing the basic structure of the plant body. Furthermore, the cells are small and protoplasm fills the cell completely. The vacuoles are extremely small. The cytoplasm does not contain differentiated plastids (chloroplasts or chromoplasts), although they are present in rudimentary form (proplastids). Meristematic cells are packed closely together without intercellular cavities. The cell wall is a very thin *primary cell wall* as well as some are thick in some plants.Maintenance of the cells requires a balance between two antagonistic processes: organ initiation and stem cell population renewal.

There are three types of meristematic tissues: *apical* (at the tips), *intercalary* (in the middle) and *lateral* (at the sides). At the meristem summit, there is a small group of slowly dividing cells, which is commonly called the central zone. Cells of this zone have a stem cell function and are essential for meristem maintenance. The proliferation and growth rates at the meristem summit usually differ considerably from those at the periphery.

The term meristem was first used in 1858 by Carl Wilhelm von Nägeli in his book Beiträge zur Wissenschaftlichen Botanik ("Contributions to Scientific Botany"). It is derived from the Greek word merizein, meaning to divide, in recognition of its inherent function.

Apical Meristems

Apical meristems are the completely undifferentiated (indeterminate) meristems in a plant. These differentiate into three kinds of primary meristems. The primary meristems in turn produce the two secondary meristem types. These secondary meristems are also known as lateral meristems because they are involved in lateral growth.

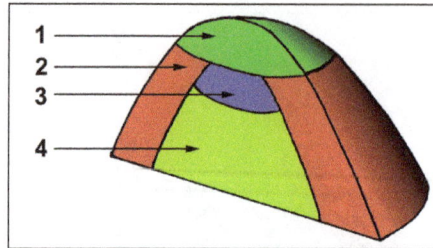

Organisation of an apical meristem (growing tip)
1 - Central zone 2 - Peripheral zone
3 - Medullary (i.e. central) meristem 4 - Medullary tissue.

There are two types of apical meristem tissue: shoot apical meristem (SAM), which gives rise to organs like the leaves and flowers, and root apical meristem (RAM), which provides the meristematic cells for future root growth. SAM and RAM cells divide rapidly and are considered indeterminate, in that they do not possess any defined end status. In that sense, the meristematic cells are frequently compared to the stem cells in animals, which have an analogous behavior and function.

The number of layers varies according to plant type. In general the outermost layer is called the tunica while the innermost layers are the corpus. In monocots, the tunica determine the physical characteristics of the leaf edge and margin. In dicots, layer two of the corpus determine the characteristics of the edge of the leaf. The corpus and tunica play a critical part of the plant physical appearance as all plant cells are formed from the meristems. Apical meristems are found in two locations: the root and the stem. Some Arctic plants have an apical meristem in the lower/middle parts of the plant. It is thought that this kind of meristem evolved because it is advantageous in Arctic conditions.

Shoot Apical Meristems

Shoot apical meristems of *Crassula ovata* (left). Fourteen days later, leaves have developed (right).

Shoot apical meristems are the source of all above-ground organs, such as leaves and flowers. Cells at the shoot apical meristem summit serve as stem cells to the surrounding peripheral region, where they proliferate rapidly and are incorporated into differentiating leaf or flower primordia.

The shoot apical meristem is the site of most of the embryogenesis in flowering plants. Primordia of leaves, sepals, petals, stamens, and ovaries are initiated here at the rate of one every time interval, called a plastochron. It is where the first indications that flower development has been evoked are manifested. One of these indications might be the loss of apical dominance and the release of otherwise dormant cells to develop as auxiliary shoot meristems, in some species in axils of primordia as close as two or three away from the apical dome. The shoot apical meristem consists of 4 distinct cell groups:

- Stem cells,

- The immediate daughter cells of the stem cells,

- A subjacent organizing center,

- Founder cells for organ initiation in surrounding regions.

The four distinct zones mentioned above are maintained by a complex signalling pathway. In *Arabidopsis thaliana*, 3 interacting *CLAVATA* genes are required to regulate the size of the stem cell reservoir in the shoot apical meristem by controlling the rate of cell division. CLV1 and CLV2 are predicted to form a receptor complex (of the LRR receptor-like kinase family) to which CLV3 is a ligand. CLV3 shares some homology with the ESR proteins of maize, with a short 14 amino acid region being conserved between the proteins. Proteins that contain these conserved regions have been grouped into the CLE family of proteins.

CLV1 has been shown to interact with several cytoplasmic proteins that are most likely involved in downstream signalling. For example, the CLV complex has been found to be associated with Rho/Rac small GTPase-related proteins. These proteins may act as an intermediate between the CLV complex and a mitogen-activated protein kinase (MAPK), which is often involved in signalling cascades. KAPP is a kinase-associated protein phosphatase that has been shown to interact with CLV1. KAPP is thought to act as a negative regulator of CLV1 by dephosphorylating it.

Another important gene in plant meristem maintenance is *WUSCHEL* (shortened to *WUS*), which is a target of CLV signaling in addition to positively regulating CLV, thus forming a feedback loop. *WUS* is expressed in the cells below the stem cells of the meristem and its presence prevents the differentiation of the stem cells. CLV1 acts to promote cellular differentiation by repressing *WUS* activity outside of the central zone containing the stem cells.

Root Apical Meristem

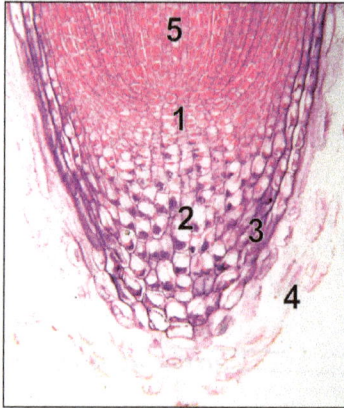

10x microscope image of root tip with meristem
1 - quiescent center 2 - calyptrogen (live rootcap cells) 3 - rootcap
4 - sloughed off dead rootcap cells 5 - procambium.

Unlike the shoot apical meristem, the root apical meristem produces cells in two dimensions. It harbors two pools of stem cells around an organizing center called the quiescent center (QC) cells and together produces most of the cells in an adult root. At its apex, the root meristem is covered by the root cap, which protects and guides its growth trajectory. Cells are continuously sloughed off the outer surface of the root cap. The QC cells are characterized by their low mitotic activity. Evidence suggests that the QC maintains the surrounding stem cells by preventing their differentiation, via signal(s) that are yet to be discovered. This allows a constant supply of new cells in the meristem required for continuous root growth. Recent findings indicate that QC can also act as a reservoir of stem cells to replenish whatever is lost or damaged. Root apical meristem and tissue patterns become established in the embryo in the case of the primary root, and in the new lateral root primordium in the case of secondary roots.

Intercalary Meristem

In angiosperms, intercalary meristems occur only in monocot (in particular, grass) stems at the base of nodes and leaf blades. Horsetails also exhibit intercalary growth. Intercalary meristems are capable of cell division, and they allow for rapid growth and regrowth of many monocots. Intercalary meristems at the nodes of bamboo allow for rapid stem elongation, while those at the base of most grass leaf blades allow damaged leaves to rapidly regrow. This leaf regrowth in grasses evolved in response to damage by grazing herbivores.

Floral Meristem

When plants begin developmental process known as flowering, the shoot apical meristem is transformed into an inflorescence meristem, which goes on to produce the floral meristem, which produces the sepals, petals, stamens, and carpels of the flower.

In contrast to vegetative apical meristems and some efflorescence meristems, floral meristems cannot continue to grow indefinitely. Their future growth is limited to the flower with a particular size and form. The transition from shoot meristem to floral meristem requires floral meristem identity genes, that both specify the floral organs and cause the termination of the production of stem cells. *AGAMOUS* (*AG*) is a floral homeotic gene required for floral meristem termination and necessary for proper development of the stamens and carpels. *AG* is necessary to prevent the conversion of floral meristems to inflorescence shoot meristems, but is identity gene *LEAFY* (*LFY*) and *WUS* and is restricted to the centre of the floral meristem or the inner two whorls. This way floral identity and region specificity is achieved. WUS activates AG by binding to a consensus sequence in the AG's second intron and LFY binds to adjacent recognition sites. Once AG is activated it represses expression of WUS leading to the termination of the meristem.

Through the years, scientists have manipulated floral meristems for economic reasons. An example is the mutant tobacco plant "Maryland Mammoth." In 1936, the department of agriculture of Switzerland performed several scientific tests with this plant. "Maryland Mammoth" is peculiar in that it grows much faster than other tobacco plants.

Apical Dominance

Apical dominance is the phenomenon where one meristem prevents or inhibits the growth of other meristems. As a result, the plant will have one clearly defined main trunk. For example, in trees, the tip of the main trunk bears the dominant shoot meristem. Therefore, the tip of the trunk grows rapidly and is not shadowed by branches. If the dominant meristem is cut off, one or more branch tips will assume dominance. The branch will start growing faster and the new growth will be vertical. Over the years, the branch may begin to look more and more like an extension of the main trunk. Often several branches will exhibit this behavior after the removal of apical meristem, leading to a bushy growth.

The mechanism of apical dominance is based on auxins, types of plant growth regulators. These are produced in the apical meristem and transported towards the roots in the cambium. If apical dominance is complete, they prevent any branches from forming as long as the apical meristem is active. If the dominance is incomplete, side branches will develop.

Recent investigations into apical dominance and the control of branching have revealed a new plant hormone family termed strigolactones. These compounds were previously known to be involved in seed germination and communication with mycorrhizal fungi and are now shown to be involved in inhibition of branching.

Diversity in Meristem Architectures

The SAM contains a population of stem cells that also produce the lateral meristems while the stem elongates. It turns out that the mechanism of regulation of the stem cell number might be evolutionarily conserved. The *CLAVATA* gene *CLV2* responsible for maintaining the stem cell population in *Arabidopsis thaliana* is very closely related to the Maize gene *FASCIATED EAR 2(FEA2)* also involved in the same function. Similarly, in Rice, the *FON1-FON2* system seems to bear a close relationship with the CLV signaling system in *Arabidopsis thaliana*. These studies suggest that the regulation of stem cell number, identity and differentiation might be an evolutionarily conserved mechanism in monocots, if not in angiosperms. Rice also contains another genetic system distinct from *FON1-FON2*, that is involved in regulating stem cell number. This example underlines the innovation that goes about in the living world all the time.

Role of the KNOX-Family Genes

Genetic screens have identified genes belonging to the KNOX family in this function. These genes essentially maintain the stem cells in an undifferentiated state. The KNOX family has undergone quite a bit of evolutionary diversification while keeping the overall mechanism more or less similar. Members of the KNOX family have been found in plants as diverse as Arabidopsis thaliana, rice, barley and tomato. KNOX-like genes are also present in some algae, mosses, ferns and gymnosperms. Misexpression of these genes leads to the formation of interesting morphological features.

The long spur of the above flower. Spurs attract pollinators and confer pollinator specificity.

Complex leaves of *C. hirsuta* result from KNOX gene expression.

For example, among members of *Antirrhineae*, only the species of the genus Antirrhinum lack a structure called spur in the floral region. A spur is considered an evolutionary innovation because it defines pollinator specificity and attraction. Researchers carried out transposon mutagenesis in *Antirrhinum majus*, and saw that some insertions led to formation of spurs that were very similar to the other members of *Antirrhineae*, indicating that the loss of spur in wild *Antirrhinum majus* populations could probably be an evolutionary innovation.

The KNOX family has also been implicated in leaf shape evolution. One study looked at the pattern of KNOX gene expression in *A. thaliana*, that has simple leaves and *Cardamine hirsuta*, a plant having complex leaves. In *A. thaliana*, the KNOX genes are completely turned off in leaves, but in *C.hirsuta*, the expression continued, generating complex leaves. Also, it has been proposed that the mechanism of KNOX gene action is conserved across all vascular plants, because there is a tight correlation between KNOX expression and a complex leaf morphology.

Primary Meristems

Apical meristems may differentiate into three kinds of primary meristem:

- Protoderm: lies around the outside of the stem and develops into the epidermis.

- Procambium: lies just inside of the protoderm and develops into primary xylem and primary phloem. It also produces the vascular cambium, and cork cambium, secondary meristems. The cork cambium further differentiates into the phelloderm (to the inside) and the phellem, or cork (to the outside). All three of these layers (cork cambium, phellem, and phelloderm) constitute the periderm. In roots, the procambium can also give rise to the pericycle, which produces lateral roots in eudicots.

- Ground meristem: develops into the cortex and the pith. Composed of parenchyma, collenchyma and sclerenchyma cells.

These meristems are responsible for primary growth, or an increase in length or height, which were discovered by scientist Joseph D. Carr of North Carolina in 1943.

Secondary Meristems

There are two types of secondary meristems, these are also called the *lateral meristems* because they surround the established stem of a plant and cause it to grow laterally (i.e., larger in diameter).

- Vascular cambium, which produces secondary xylem and secondary phloem. This is a process that may continue throughout the life of the plant. This is what gives rise to wood in plants. Such plants are called arborescent. This does not occur in plants that do not go through secondary growth (known as herbaceous plants).

- Cork cambium, which gives rise to the periderm, which replaces the epidermis.

Indeterminate Growth of Meristems

Though each plant grows according to a certain set of rules, each new root and shoot meristem can go on growing for as long as it is alive. In many plants, meristematic growth is potentially indeterminate, making the overall shape of the plant not determinate in advance. This is the primary growth. Primary growth leads to lengthening of the plant body and organ formation. All plant organs arise ultimately from cell divisions in the apical meristems, followed by cell expansion and differentiation. Primary growth gives rise to the apical part of many plants.

The growth of nitrogen-fixing root nodules on legume plants such as soybean and pea is either determinate or indeterminate. Thus, soybean (or bean and Lotus japonicus) produce determinate nodules (spherical), with a branched vascular system surrounding the central infected zone. Often, Rhizobium infected cells have only small vacuoles. In contrast, nodules on pea, clovers, and Medicago truncatula are indeterminate, to maintain (at least for some time) an active meristem that yields new cells for Rhizobium infection. Thus zones of maturity exist in the nodule. Infected cells usually possess a large vacuole. The plant vascular system is branched and peripheral.

Cloning

Under appropriate conditions, each shoot meristem can develop into a complete, new plant or clone. Such new plants can be grown from shoot cuttings that contain an apical meristem. Root apical meristems are not readily cloned, however. This cloning is called asexual reproduction or vegetative reproduction and is widely practiced in horticulture to mass-produce plants of a desirable genotype. This process is also known as mericloning.

Propagating through cuttings is another form of vegetative propagation that initiates root or shoot production from secondary meristematic cambial cells. This explains why basal 'wounding' of shoot-borne cuttings often aids root formation.

Induced Meristems

Meristems may also be induced in the roots of legumes such as soybean, *Lotus japonicus*, pea, and *Medicago truncatula* after infection with soil bacteria commonly called Rhizobia.Cells of the inner or outer cortex in the so-called "window of nodulation" just behind the developing root tip are induced to divide. The critical signal substance is the lipo-oligosaccharide Nod factor, decorated with side groups to allow specificity of interaction. The Nod factor receptor proteins NFR1 and NFR5 were cloned from several legumes including *Lotus japonicus, Medicago truncatula* and soybean (*Glycine max*). Regulation of nodule meristems utilizes long-distance regulation known as the autoregulation of nodulation (AON). This process involves a leaf-vascular tissue located LRR receptor kinases (LjHAR1, GmNARK and MtSUNN), CLE peptide signalling, and KAPP interaction, similar to that seen in the CLV1,2,3 system. LjKLAVIER also exhibits a nodule regulation phenotype though it is not yet known how this relates to the other AON receptor kinases.

References

- Dickison, W.C. (2000). Integrative Plant Anatomy (page 196). Elsevier Science. ISBN 9780080508917. Archived from the original on 2017-11-06

- Plant-tissue, plant-kingdom: tutorvista.com, Retrieved 30 January, 2019

- Keith Roberts, ed. (2007). Handbook of Plant Science. 1 (illustrated ed.). John Wiley & Sons. P. 185. ISBN 9780470057230

- Lalonde S. Wipf D., Frommer W.B. (2004). "Transport mechanisms for organic forms of carbon and nitrogen between source and sink". Annu Rev Plant Biol. 55: 341–72. Doi:10.1146/annurev. arplant.55.031903.141758. PMID 15377224

- Parenchyma-plant-tissue, science: britannica.com, Retrieved 19 April, 2019

- Campbell, Neil A.; Reece, Jane B. (2008). Biology (8th ed.). Pearson Education, Inc. Pp. 744–745. ISBN 978-0-321-54325-7

- Collenchyma, science: britannica.com, Retrieved 25 February, 2019

- Fletcher, J. C. (2002). "Shoot and Floral Meristem Maintenance in Arabidopsis". Annu. Rev. Plant Biol. 53: 45–66. Doi:10.1146/annurev.arplant.53.092701.143332. PMID 12221985

- Sclerenchyma, science: britannica.com, Retrieved 26 July, 2019

- Valster, A. H.; et al. (2000). "Plant gtpases: the Rhos in bloom". Trends in Cell Biology. 10 (4): 141–146. Doi:10.1016/s0962-8924(00)01728-1

- W.S. Judd, C.S Campbell, E.A. Kellogg, P.F. Stevens, M.J. Donoghue (2002). Plant systematics: A phylogenetic approach (2 ed.). Sinauer Associates Inc. ISBN 0-87893-403-0

Plant Anatomy and Physiology

The study of the internal structure of plants is referred to as plant anatomy. The sub-discipline of botany that is concerned with the functioning and physiology of plants is known as plant physiology. This chapter has been carefully written to provide an easy understanding of plant anatomy and plant physiology.

PLANT ANATOMY

Plant anatomy is the study of plant tissues and cells in order to learn more about the way these organisms are constructed and how they work. These studies are very important because they lead to a better understanding of how to care for plants and fight plant diseases. Plant anatomy is also known as phytotomy.

A plant is a complex structure that consists of a number of parts which constitute the whole plant. If you learn to identify each individual part, you will gain a much greater understanding as to how the plant works as a whole. This can be helpful to aromatherapists who need to be aware of the part of the plant an essential oil was derived from because there is often a connection between the oils location in a plant and its therapeutic action. Understanding plant anatomy also helps everyone appreciate the art of distillation and extraction.

The Life Span of a Plant

Plant species vary not only in appearance, but also in their longevity (length of life). Annual flowering plants only live for one year whereas biennial plants live for two years, producing only leaves in the first year and flowering in the next. Perennial plants live for more than two years. They might be evergreen (never losing their leaves) or deciduous (loses the leaves in autumn). A plant's status as an annual, biennial, or perennial can vary due to its geographical location and purpose of cultivation. Essential oils are extracted from a mixture of these different biological lifecycles. All plants consist of some basic parts as follows:

The Flower

Not all plants flower but many plants from which essential oils are extracted from are flowering plants; for example, lavender (Lavandula angustifolia), rose (Rosa damascena) and rosemary (Rosmarinus officinalis).

The flower of a plant is a complex structure. These are the various parts which make up the flower of a plant:

- The petals (made up of the corolla).

- The calyx (the outer, or green, leaves).

- The stamen (containing the pollen which insects and birds are attracted to).

- The pistil (containing the ovary, the style and the stigma of the flower).

The Fruits and Seeds

The seed of a plant contains the nucleus; a new plant grows from the seed as long as the growing conditions are right for it to do so. Plants also contain fruits which might be described in one of the following ways:

- Follicule

- Legume (pod)

- Drupe

- Achenium

- Caryopsis

- Cremocarp

- Nut

- Berry

- Samara

- Pome

- Pepo

- Silique

- Capsule

- Cone

Plants that have fruits from which an essential oil is extracted include lemon (Citrus limon) and sweet orange (Citrus sinensis).

The Leaves

The leaves grow on the part of the stalk called the petiole. Leaves might be short, fat, long, thin, hairy, curvy, indented, wispy or any number of other shapes, textures and colors. The various types of leaves on a plant are botanically identified as follows:

- Lanceolate
- Cuneiform
- Sagittate
- Ovate
- Cordate
- Pinnate
- Pectinate
- Runcinate
- Lyrate
- Palmate
- Pedate
- Obovate
- Reniform
- Hastate
- Serrate
- Peltate
- Dentate
- Crenate
- Sinuate

Plants that produce a leaf essential oil include cinnamon (Cinnamomum zeylanicum) and petitgrain (Citrus aurantium var. amara).

The Stem

Stems are found on all flowering plants and gravitate towards the light and air, away from the root. Some plants might appear stemless but they actually have the stem below ground or the stem is extremely short. A tree's stem is better known as the trunk. Herbs have stems which die after flowering. Essential oils are extracted from all of these types of plants.

Clove (Syzygium aromaticum) produces a stem essential oil, although clove bud is always the more preferable essential oil for aromatherapists to use since it is much less irritant to the skin.

The Roots

The root of a plant is usually located in the soil below the plant. It acts as an anchor for the plant. Types of roots include:

- Fusiform root: Root tapers both up and down, for example, a radish (Rhapanus sativus).

- Fasciculated root: The fibers or branches are thickened.

- Tuberiferous root: Some of the branches of the root become rounded knobs, such as in a potato (Solanum tuberosum) and sometimes culminate in a branch known as a palmate root.

- Aerial root: The root actually grows into the open air, such as in Indian Corn.

- Conical root: The root tapers regularly from the crown to the apex of the plant, for example, a carrot (Daucus carota).

- Napiform root: The root is swollen at the base and extends horizontally more than vertically, such as in a turnip (Brassica napa).

- Rhizome root: Thick, spreading root such as in ginger (Zingiber officinale).

Ginger (Zingiber officinale) produces an essential oil from the roots of the plant.

PLANT PHYSIOLOGY

Plant physiology is a subdiscipline of botany concerned with the functioning, or physiology, of plants. Closely related fields include plant morphology (structure of plants), plant ecology (interactions with the environment), phytochemistry (biochemistry of plants), cell biology, genetics, biophysics and molecular biology.

Fundamental processes such as photosynthesis, respiration, plant nutrition, plant hormone functions, tropisms, nastic movements, photoperiodism, photomorphogenesis, circadian rhythms, environmental stress physiology, seed germination, dormancy and stomata function and transpiration, both parts of plant water relations, are studied by plant physiologists.

A germination rate experiment.

Aims

The field of plant physiology includes the study of all the internal activities of plants—those chemical and physical processes associated with life as they occur in plants. This includes study at many levels of scale of size and time. At the smallest scale are molecular interactions of photosynthesis and internal diffusion of water, minerals, and nutrients. At the largest scale are the processes of plant development, seasonality, dormancy, and reproductive control. Major subdisciplines of plant physiology include phytochemistry (the study of the biochemistry of plants) and phytopathology (the study of disease in plants). The scope of plant physiology as a discipline may be divided into several major areas of research.

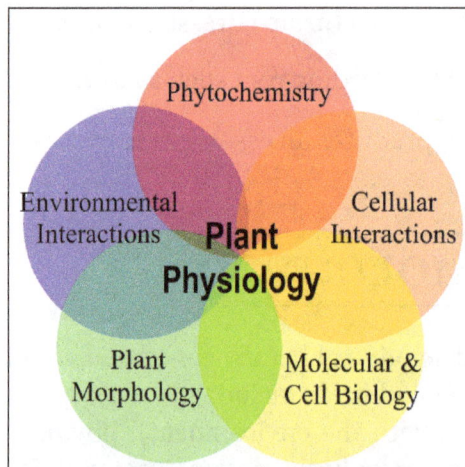
Five key areas of study within plant physiology.

First, the study of phytochemistry (plant chemistry) is included within the domain of plant physiology. To function and survive, plants produce a wide array of chemical compounds not found in other organisms. Photosynthesis requires a large array of pigments, enzymes, and other compounds to function. Because they cannot move, plants must also defend themselves chemically from herbivores, pathogens and competition from other plants. They do this by producing toxins and foul-tasting or smelling chemicals. Other compounds defend plants against disease, permit survival during drought, and prepare plants for dormancy, while other compounds are used to attract pollinators or herbivores to spread ripe seeds.

Secondly, plant physiology includes the study of biological and chemical processes of individual plant cells. Plant cells have a number of features that distinguish them from cells of animals, and which lead to major differences in the way that plant life behaves and responds differently from animal life. For example, plant cells have a cell wall which restricts the shape of plant cells and thereby limits the flexibility and mobility of plants. Plant cells also contain chlorophyll, a chemical compound that interacts with light in a way that enables plants to manufacture their own nutrients rather than consuming other living things as animals do.

Thirdly, plant physiology deals with interactions between cells, tissues, and organs within a plant. Different cells and tissues are physically and chemically specialized to perform different functions. Roots and rhizoids function to anchor the plant and acquire minerals in the soil. Leaves catch light in order to manufacture nutrients. For both of these organs to remain living, minerals that the roots acquire must be transported to the leaves, and the nutrients manufactured in the leaves must be transported to the roots. Plants have developed a number of ways to achieve this transport, such as vascular tissue, and the functioning of the various modes of transport is studied by plant physiologists.

Fourthly, plant physiologists study the ways that plants control or regulate internal functions. Like animals, plants produce chemicals called hormones which are produced in one part of the plant to signal cells in another part of the plant to respond. Many flowering plants bloom at the appropriate time because of light-sensitive compounds that respond to the length of the night, a phenomenon known as photoperiodism. The ripening of fruit and loss of leaves in the winter are controlled in part by the production of the gas ethylene by the plant.

Finally, plant physiology includes the study of plant response to environmental conditions and their variation, a field known as environmental physiology. Stress from water loss, changes in air chemistry, or crowding by other plants can lead to changes in the way a plant functions. These changes may be affected by genetic, chemical, and physical factors.

Biochemistry of Plants

The chemical elements of which plants are constructed—principally carbon, oxygen, hydrogen, nitrogen, phosphorus, sulfur, etc.—are the same as for all other life forms animals, fungi, bacteria and even viruses. Only the details of the molecules into which they are assembled differs.

Latex being collected from a tapped rubber tree.

Despite this underlying similarity, plants produce a vast array of chemical compounds with unique properties which they use to cope with their environment. Pigments are used by plants to absorb or detect light, and are extracted by humans for use in dyes. Other plant products may be used for the manufacture of commercially important rubber or biofuel. Perhaps the most celebrated compounds from plants are those with pharmacological activity, such as salicylic acid from which aspirin is made, morphine, and digoxin. Drug companies spend billions of dollars each year researching plant compounds for potential medicinal benefits.

Constituent Elements

Plants require some nutrients, such as carbon and nitrogen, in large quantities to survive. Some nutrients are termed macronutrients, where the prefix macro- (large) refers to the quantity needed, not the size of the nutrient particles themselves. Other nutrients, called micronutrients, are required only in trace amounts for plants to remain healthy. Such micronutrients are usually absorbed as ions dissolved in water taken from the soil, though carnivorous plants acquire some of their micronutrients from captured prey.

The following tables list element nutrients essential to plants. Uses within plants are generalized.

Macronutrients – Necessary in Large Quantities		
Element	Form of uptake	Notes
Nitrogen	NO_3^-, NH_4^+	Nucleic acids, proteins, hormones, etc.
Oxygen	O_2, H_2O	Cellulose, starch, other organic compounds
Carbon	CO_2	Cellulose, starch, other organic compounds
Hydrogen	H_2O	Cellulose, starch, other organic compounds
Potassium	K^+	Cofactor in protein synthesis, water balance, etc.

Calcium	Ca^{2+}	Membrane synthesis and stabilization
Magnesium	Mg^{2+}	Element essential for chlorophyll
Phosphorus	$H_2PO_4^-$	Nucleic acids, phospholipids, ATP
Sulfur	SO_4^{2-}	Constituent of proteins

Micronutrients – Necessary in Small Quantities		
Element	Form of uptake	Notes
Chlorine	Cl^-	Photosystem II and stomata function
Iron	Fe^{2+}, Fe^{3+}	Chlorophyll formation and nitrogen fixation
Boron	HBO_3	Crosslinking pectin
Manganese	Mn^{2+}	Activity of some enzymes and photosystem II
Zinc	Zn^{2+}	Involved in the synthesis of enzymes and chlorophyll
Copper	Cu^+	Enzymes for lignin synthesis
Molybdenum	MoO_4^{2-}	Nitrogen fixation, reduction of nitrates
Nickel	Ni^{2+}	Enzymatic cofactor in the metabolism of nitrogen compounds

Pigments

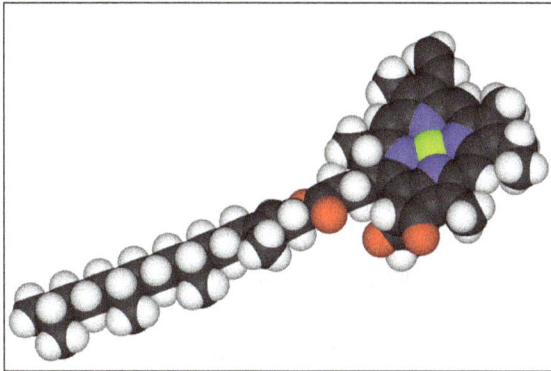

Space-filling model of the chlorophyll molecule.

Among the most important molecules for plant function are the pigments. Plant pigments include a variety of different kinds of molecules, including porphyrins, carotenoids, and anthocyanins. All biological pigments selectively absorb certain wavelengths of light while reflecting others. The light that is absorbed may be used by the plant to power chemical reactions, while the reflected wavelengths of light determine the color the pigment appears to the eye.

Chlorophyll is the primary pigment in plants; it is a porphyrin that absorbs red and blue wavelengths of light while reflecting green. It is the presence and relative abundance of chlorophyll that gives plants their green color. All land plants and green algae possess two forms of this pigment: chlorophyll a and chlorophyll b. Kelps, diatoms, and other photosynthetic heterokonts contain chlorophyll c instead of b, red algae possess chlorophyll a. All chlorophylls serve as the primary means plants use to intercept light to fuel photosynthesis.

Anthocyanin gives these pansies their dark purple pigmentation.

Carotenoids are red, orange, or yellow tetraterpenoids. They function as accessory pigments in plants, helping to fuel photosynthesis by gathering wavelengths of light not readily absorbed by chlorophyll. The most familiar carotenoids are carotene (an orange pigment found in carrots), lutein (a yellow pigment found in fruits and vegetables), and lycopene (the red pigment responsible for the color of tomatoes). Carotenoids have been shown to act as antioxidants and to promote healthy eyesight in humans.

Anthocyanins (literally "flower blue") are water-soluble flavonoid pigments that appear red to blue, according to pH. They occur in all tissues of higher plants, providing color in leaves, stems, roots, flowers, and fruits, though not always in sufficient quantities to be noticeable. Anthocyanins are most visible in the petals of flowers, where they may make up as much as 30% of the dry weight of the tissue. They are also responsible for the purple color seen on the underside of tropical shade plants such as Tradescantia zebrina. In these plants, the anthocyanin catches light that has passed through the leaf and reflects it back towards regions bearing chlorophyll, in order to maximize the use of available light.

Betalains are red or yellow pigments. Like anthocyanins they are water-soluble, but unlike anthocyanins they are indole-derived compounds synthesized from tyrosine. This class of pigments is found only in the Caryophyllales (including cactus and amaranth), and never co-occur in plants with anthocyanins. Betalains are responsible for the deep red color of beets, and are used commercially as food-coloring agents. Plant physiologists are uncertain of the function that betalains have in plants which possess them, but there is some preliminary evidence that they may have fungicidal properties.

Signals and Regulators

Plants produce hormones and other growth regulators which act to signal a physiological response in their tissues. They also produce compounds such as phytochrome that are sensitive to light and which serve to trigger growth or development in response to environmental signals.

A mutation that stops Arabidopsis thaliana responding
to auxin causes abnormal growth (right).

Plant Hormones

Plant hormones, known as plant growth regulators (PGRs) or phytohormones, are chemicals that regulate a plant's growth. According to a standard animal definition, hormones are signal molecules produced at specific locations, that occur in very low concentrations, and cause altered processes in target cells at other locations. Unlike animals, plants lack specific hormone-producing tissues or organs. Plant hormones are often not transported to other parts of the plant and production is not limited to specific locations.

Plant hormones are chemicals that in small amounts promote and influence the growth, development and differentiation of cells and tissues. Hormones are vital to plant growth; affecting processes in plants from flowering to seed development, dormancy, and germination. They regulate which tissues grow upwards and which grow downwards, leaf formation and stem growth, fruit development and ripening, as well as leaf abscission and even plant death.

The most important plant hormones are abscissic acid (ABA), auxins, ethylene, gibberellins, and cytokinins, though there are many other substances that serve to regulate plant physiology.

Photomorphogenesis

While most people know that light is important for photosynthesis in plants, few realize that plant sensitivity to light plays a role in the control of plant structural development (morphogenesis). The use of light to control structural development is called photomorphogenesis, and is dependent upon the presence of specialized photoreceptors, which are chemical pigments capable of absorbing specific wavelengths of light.

Plants use four kinds of photoreceptors: phytochrome, cryptochrome, a UV-B photoreceptor, and protochlorophyllide a. The first two of these, phytochrome and cryptochrome, are photoreceptor proteins, complex molecular structures formed by joining a protein with a light-sensitive pigment. Cryptochrome is also known as the UV-A photoreceptor, because it absorbs ultraviolet light in the long wave "A" region. The UV-B receptor is one or more compounds not yet identified with certainty, though some evidence suggests carotene or riboflavin as candidates. Protochlorophyllide a, as its name suggests, is a chemical precursor of chlorophyll.

The most studied of the photoreceptors in plants is phytochrome. It is sensitive to light in the red and far-red region of the visible spectrum. Many flowering plants use it to regulate the time of flowering based on the length of day and night (photoperiodism) and to set circadian rhythms. It also regulates other responses including the germination of seeds, elongation of seedlings, the size, shape and number of leaves, the synthesis of chlorophyll, and the straightening of the epicotyl or hypocotyl hook of dicot seedlings.

Photoperiodism

Many flowering plants use the pigment phytochrome to sense seasonal changes in day length, which they take as signals to flower. This sensitivity to day length is termed photoperiodism. Broadly speaking, flowering plants can be classified as long day plants, short day plants, or day neutral plants, depending on their particular response to changes in day length. Long day plants require a certain minimum length of daylight to starts flowering, so these plants flower in the spring or summer. Conversely, short day plants flower when the length of daylight falls below a certain critical level. Day neutral plants flower when the length of daylight falls below a certain critical level. Day neutral plants do not initiate flowering based on photoperiodism, though some may use temperature sensitivity (vernalization) instead.

The poinsettia is a short-day plant, requiring two
months of long nights prior to blooming.

Although a short day plant cannot flower during the long days of summer, it is not actually the period of light exposure that limits flowering. Rather, a short day plant requires a minimal length of uninterrupted darkness in each 24-hour period (a short daylength) before floral development can begin. It has been determined experimentally that a short day plant (long night) does not flower if a flash of phytochrome activating light is used on the plant during the night.

Plants make use of the phytochrome system to sense day length or photoperiod. This fact is utilized by florists and greenhouse gardeners to control and even induce flowering out of season, such as the Poinsettia.

Environmental Physiology

Paradoxically, the subdiscipline of environmental physiology is on the one hand a recent field of study in plant ecology and on the other hand one of the oldest. Environmental physiology is the preferred name of the subdiscipline among plant physiologists, but it goes by a number of other names in the applied sciences. It is roughly synonymous with ecophysiology, crop ecology, horticulture and agronomy. The particular name applied to the subdiscipline is specific to the viewpoint and goals of research. Whatever name is applied, it deals with the ways in which plants respond to their environment and so overlaps with the field of ecology.

Environmental physiologists examine plant response to physical factors such as radiation (including light and ultraviolet radiation), temperature, fire, and wind. Of particular importance are water relations (which can be measured with the Pressure bomb) and the stress of drought or inundation, exchange of gases with the atmosphere, as well as the cycling of nutrients such as nitrogen and carbon.

Phototropism in *Arabidopsis thaliana* is
regulated by blue to UV light.

Environmental physiologists also examine plant response to biological factors. This includes not only negative interactions, such as competition, herbivory, disease and parasitism, but also positive interactions, such as mutualism and pollination.

Tropisms and Nastic Movements

Plants may respond both to directional and non-directional stimuli. A response to a directional stimulus, such as gravity or sunlight, is called a tropism. A response to a nondirectional stimulus, such as temperature or humidity, is a nastic movement.

Tropisms in plants are the result of differential cell growth, in which the cells on one side of the plant elongates more than those on the other side, causing the part to bend toward the side with less growth. Among the common tropisms seen in plants is phototropism, the bending of the plant toward a source of light. Phototropism allows the plant to maximize light exposure in plants which require additional light for photosynthesis, or to minimize it in plants subjected to intense light and heat. Geotropism allows the roots of a plant to determine the direction of gravity and grow downwards. Tropisms generally result from an interaction between the environment and production of one or more plant hormones.

Nastic movements results from differential cell growth (e.g. epinasty and hiponasty), or from changes in turgor pressure within plant tissues (e.g., nyctinasty), which may occur rapidly. A familiar example is thigmonasty (response to touch) in the Venus fly trap, a carnivorous plant. The traps consist of modified leaf blades which bear sensitive trigger hairs. When the hairs are touched by an insect or other animal, the leaf folds shut. This mechanism allows the plant to trap and digest small insects for additional nutrients. Although the trap is rapidly shut by changes in internal cell pressures, the leaf must grow slowly to reset for a second opportunity to trap insects.

Plant Disease

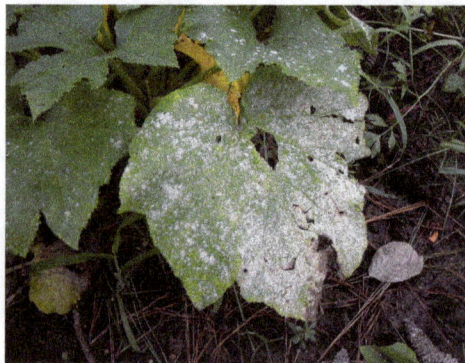

Powdery mildew on crop leaves

Economically, one of the most important areas of research in environmental physiology is that of phytopathology, the study of diseases in plants and the manner in which plants resist or cope with infection. Plant are susceptible to the same kinds of disease organisms as animals, including viruses, bacteria, and fungi, as well as physical invasion by insects and roundworms.

Because the biology of plants differs with animals, their symptoms and responses are quite different. In some cases, a plant can simply shed infected leaves or flowers to prevent the spread of disease, in a process called abscission. Most animals do not have this option as a means of controlling disease. Plant diseases organisms themselves also differ from those causing disease in animals because plants cannot usually spread infection through casual physical contact. Plant pathogens tend to spread via spores or are carried by animal vectors.

One of the most important advances in the control of plant disease was the discovery of Bordeaux mixture in the nineteenth century. The mixture is the first known fungicide and is a combination of copper sulfate and lime. Application of the mixture served to inhibit the growth of downy mildew that threatened to seriously damage the French wine industry.

Economic Applications

Food Production

In horticulture and agriculture along with food science, plant physiology is an important topic relating to fruits, vegetables, and other consumable parts of plants. Topics studied include: climatic requirements, fruit drop, nutrition, ripening, fruit set. The production of food crops also hinges on the study of plant physiology covering such topics as optimal planting and harvesting times and post harvest storage of plant products for human consumption and the production of secondary products like drugs and cosmetics.

PLANT ORGAN SYSTEM

A plant has two organ systems: 1) the shoot system, and 2) the root system. The shoot system is above ground and includes the organs such as leaves, buds, stems, flowers (if the plant has any), and fruits (if the plant has any). The root system includes those parts of the plant below ground, such as the roots, tubers, and rhizomes.

THE SHOOT SYSTEM

A plant has many complicated and complex systems that keep it living and growing, including the shoot system. When referring to the shoot system in a plant, we generally refer to the leaves, buds, flowering stems and flowering buds, as well as the main stem itself. The word 'shoot' generally is used when talking about the main stem.

As we move from the ground surface to the terminal bud (end of the undeveloped shoot) we will encounter nodes and internodes. Nodes are the points where leaves are attached, and internodes are the places on the stem between the nodes. In the crux created by the node and stem, there are axillary buds that lay dormant but have the potential to grow a vegetative branch. These axillary buds lay dormant because of apical dominance. This is a phenomena where the plant concentrates most of its resources at the terminal bud. The terminal buds grows at the apex, or tip of the plant. This is responsible for making the plant grow taller and bigger so it makes sense that the plant would want to concentrate resources here. If the terminal bud gets damaged, the axillary buds will 'wake up' and begin to grow, saving the plant.

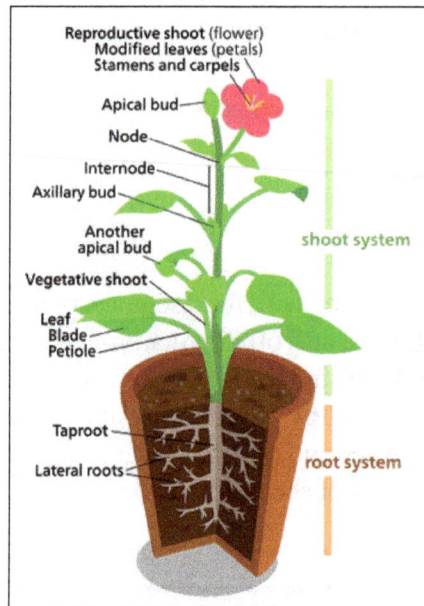

Plant Shoot System.

If the axillary buds begin to grow, they can create their own terminal buds, leaves, etc. They, in a sense, create a secondary plant themselves. If you own any houseplants, you might understand the idea of pruning them to create 'bushy' plants. Well, this is because you are removing the terminal bud that stimulates the axillary buds to grow, creating that thicker looking plant.

As is commonly known, the leaves of the plant act as the photosynthetic factory for it, producing sugars and other compounds for the plant to survive on. There are many different types of leaves, but most will include a blade, stalk, veins and the petiole. The stalk is the main stem of the leaf, the blade is the actual green leaf portion, and the petiole is where it attaches to the node. The veins are similar to our veins in that they carry water and nutrients out into the leaf.

The Leaf

The diversity of leaves.

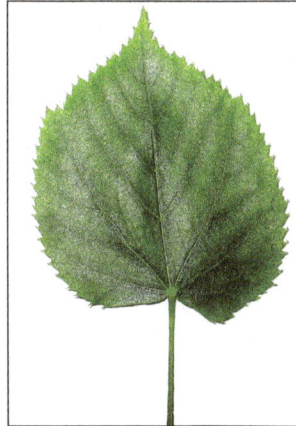

Leaf of Tilia tomentosa (Silver lime tree).

A leaf is an organ of a vascular plant and is the principal lateral appendage of the stem. The leaves and stem together form the shoot. Leaves are collectively referred to as foliage, as in "autumn foliage".

A leaf is a dorsiventrally flattened organ, usually borne above ground and specialized for photosynthesis. In most leaves, the primary photosynthetic tissue, the palisade mesophyll, is located on the upper side of the blade or lamina of the leaf but in some species, including the mature foliage of *Eucalyptus*, palisade mesophyll is present on both sides and the leaves are said to be isobilateral. Most leaves have distinct upper surface (*adaxial*) and lower surface (*abaxial*) that differ in colour, hairiness, the number of stomata (pores that intake and output gases), the amount and structure of epicuticular wax and other features.

Leaves can have many different shapes, sizes, and textures. The broad, flat leaves with complex venation of flowering plants are known as *megaphylls* and the species that bear them, the majority, as broad-leaved or megaphyllous plants. In the clubmosses, with different evolutionary origins, the leaves are simple (with only a single vein) and are known as microphylls. Some leaves, such as bulb scales, are not above ground. In many aquatic species the leaves are submerged in water. Succulent plants often have thick juicy leaves, but some leaves are without major photosynthetic function and may be dead at maturity, as in some cataphylls and spines. Furthermore, several kinds of leaf-like structures found in vascular plants are not totally homologous with them. Examples include flattened plant stems called phylloclades and cladodes, and flattened leaf stems called phyllodes which differ from leaves both in their structure and origin. Some structures of non-vascular plants look and function much like leaves. Examples include the phyllids of mosses and liverworts.

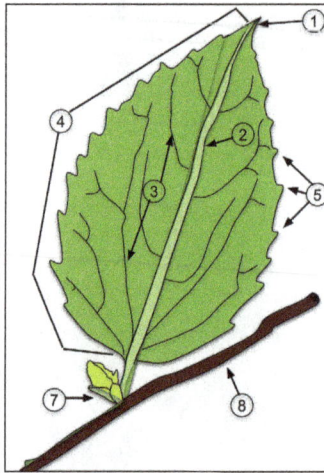

Diagram of a simple leaf.
1. Apex, 2. Midvein (Primary vein), 3. Secondary vein, 4. Lamina,
5. Leaf margin, 6. Petiole, 7. Bud, 8. Stem.

Top and Right: Staghorn Sumac, *Rhus typhina* (Compound Leaf)
Bottom: Skunk Cabbage, *Symplocarpus foetidus* (Simple Leaf)
1 Apex, 2. Primary Vein, 3. Secondary Vein, 4. Lamina, 5. Leaf Margin, 6. Petiole.

Leaves are the most important organs of most vascular plants. Green plants are autotrophic, meaning that they do not obtain food from other living things but instead create their own food by photosynthesis. They capture the energy in sunlight and use it to make simple sugars, such as glucose and sucrose, from carbon dioxide and water. The sugars are then stored as starch, further processed by chemical synthesis into more complex organic molecules such as proteins or cellulose, the basic structural material in plant cell walls, or metabolised by cellular respiration to provide chemical energy to run cellular processes. The leaves draw water from the ground in the transpiration stream through a vascular conducting system known as xylem and obtain carbon dioxide from the atmosphere by diffusion through openings called stomata in the outer covering layer of the leaf (epidermis), while leaves are orientated to maximise their exposure to sunlight. Once sugar has been synthesized, it needs to be transported to areas of active growth such as the plant shoots and roots. Vascular plants transport sucrose in a special tissue called the phloem. The phloem and xylem are parallel to each other but the transport of materials is usually in opposite directions. Within the leaf these vascular systems branch (ramify) to form veins which supply as much of the leaf as possible, ensuring that cells carrying out photosynthesis are close to the transportation system.

3D rendering of a computed tomography scan of a leaf.

Typically leaves are broad, flat and thin (dorsiventrally flattened), thereby maximising the surface area directly exposed to light and enabling the light to penetrate the tissues and reach the chloroplasts, thus promoting photosynthesis. They are arranged on the plant so as to expose their surfaces to light as efficiently as possible without shading each other, but there are many exceptions and complications. For instance plants adapted to windy conditions may have pendent leaves, such as in many willows and eucalyptss. The flat, or laminar, shape also maximises thermal contact with the surrounding air, promoting cooling. Functionally, in addition to carrying out photosynthesis, the leaf is the principal site of transpiration, providing the energy required to draw the transpiration stream up from the roots, and guttation.

Many gymnosperms have thin needle-like or scale-like leaves that can be advantageous in cold climates with frequent snow and frost. These are interpreted as reduced from megaphyllous leaves of their Devonian ancestors. Some leaf forms are adapted to modulate the amount of light they absorb to avoid or mitigate excessive heat, ultraviolet damage, or desiccation, or to sacrifice light-absorption efficiency in favour of protection from herbivory. For xerophytes the major constraint is not light flux or intensity, but drought. Some window plants such as *Fenestraria* species and some *Haworthia* species such as *Haworthia tesselata* and *Haworthia truncata* are examples of xerophytes. and Bulbine mesembryanthemoides.

Leaves also function to store chemical energy and water (especially in succulents) and may become specialised organs serving other functions, such as tendrils of peas and other legumes, the protective spines of cacti and the insect traps in carnivorous plants such as *Nepenthes* and *Sarracenia*. Leaves are the fundamental structural units from which cones are constructed in gymnosperms (each cone scale is a modified megaphyll leaf known as a sporophyll) and from which flowers are constructed in flowering plants.

Vein skeleton of a leaf. Veins contain lignin that make them harder to degrade for microorganisms.

The internal organisation of most kinds of leaves has evolved to maximise exposure of the photosynthetic organelles, the chloroplasts, to light and to increase the absorption of carbon dioxide while at the same time controlling water loss. Their surfaces are waterproofed by the plant cuticle and gas exchange between the mesophyll cells and the atmosphere is controlled by minute (length and width measured in tens of μm) openings called stomata which open or close to regulate the rate exchange of carbon dioxide, oxygen, and water vapour into and out of the internal intercellular space system. Stomatal opening is controlled by the turgor pressure in a pair of guard cells that surround the stomatal aperture. In any square centimeter of a plant leaf there may be from 1,000 to 100,000 stomata.

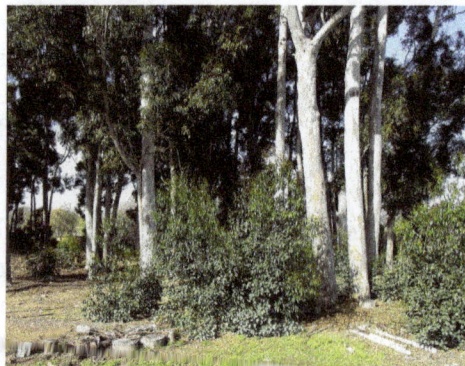

Near the ground these *Eucalyptus* saplings have juvenile dorsiventral foliage from the previous year, but this season their newly sprouting foliage is isobilateral, like the mature foliage on the adult trees above.

The shape and structure of leaves vary considerably from species to species of plant, depending largely on their adaptation to climate and available light, but also to other factors such as grazing animals (such as deer), available nutrients, and ecological competition from other plants. Considerable changes in leaf type occur within species too, for example as a plant matures; as a case in point *Eucalyptus* species commonly have isobilateral, pendent leaves when mature and dominating their neighbours; however, such trees tend to have erect or horizontal dorsiventral leaves as seedlings, when their growth is limited by the available light. Other factors include the need to balance water loss at high temperature and low humidity against the need to absorb atmospheric carbon dioxide. In most plants leaves also are the primary organs responsible for transpiration and guttation (beads of fluid forming at leaf margins).

Leaves can also store food and water, and are modified accordingly to meet these functions, for example in the leaves of succulent plants and in bulb scales. The concentration of photosynthetic structures in leaves requires that they be richer in protein, minerals, and sugars than, say, woody stem tissues. Accordingly, leaves are prominent in the diet of many animals.

A leaf shed in autumn.

Correspondingly, leaves represent heavy investment on the part of the plants bearing them, and their retention or disposition are the subject of elaborate strategies for dealing with pest pressures, seasonal conditions, and protective measures such as the growth of thorns and the production of phytoliths, lignins, tannins and poisons.

Deciduous plants in frigid or cold temperate regions typically shed their leaves in autumn, whereas in areas with a severe dry season, some plants may shed their leaves until the dry season ends. In either case the shed leaves may be expected to contribute their retained nutrients to the soil where they fall.

In contrast, many other non-seasonal plants, such as palms and conifers, retain their leaves for long periods; *Welwitschia* retains its two main leaves throughout a lifetime that may exceed a thousand years.

The leaf-like organs of Bryophytes (e.g., mosses and liverworts), known as phyllids, differ morphologically from the leaves of vascular plants in that they lack vascular tissue, are usually only a single cell thick and have no cuticle stomata or internal system of intercellular spaces.

Simple, vascularised leaves (microphylls) first evolved as enations, extensions of the stem, in clubmosses such as *Baragwanathia* during the Silurian period. True leaves or euphylls of larger size and with more complex venation did not become widespread in other groups until the Devonian period, by which time the carbon dioxide concentration in the atmosphere had dropped significantly. This occurred independently in several separate lineages of vascular plants, in progymnosperms like *Archaeopteris*, in Sphenopsida, ferns and later in the gymnosperms and angiosperms. Euphylls are also referred to as macrophylls or megaphylls (large leaves).

Morphology (Large-Scale Features)

A structurally complete leaf of an angiosperm consists of a petiole (leaf stalk), a lamina (leaf blade), and stipules (small structures located to either side of the base of the petiole). Not every species produces leaves with all of these structural components. Stipules may be conspicuous (e.g. beans and roses), soon falling or otherwise not obvious as in Moraceae or absent altogether as in the Magnoliaceae. A petiole may be absent, or the blade may not be laminar (flattened). The tremendous variety shown in leaf structure (anatomy) from species to species is presented in detail below under morphology. The petiole mechanically links the leaf to the plant and provides the route for transfer of water and sugars to and from the leaf. The lamina is typically the location of the majority of photosynthesis. The upper (adaxial) angle between a leaf and a stem is known as the axil of the leaf. It is often the location of a bud. Structures located there are called "axillary".

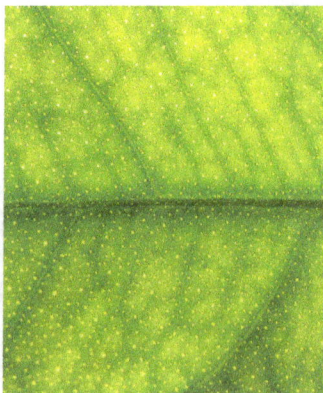

Translucent glands in *Citrus* leaves.

External leaf characteristics, such as shape, margin, hairs, the petiole, and the presence of stipules and glands, are frequently important for identifying plants to family, genus or species levels, and botanists have developed a rich terminology for describing leaf characteristics. Leaves almost always have determinate growth. They grow to a specific pattern and shape and then stop. Other plant parts like stems or roots have non-determinate growth, and will usually continue to grow as long as they have the resources to do so.

The type of leaf is usually characteristic of a species (monomorphic), although some species produce more than one type of leaf (dimorphic or polymorphic). The longest leaves are those of the Raffia palm, *R. regalis* which may be up to 25 m (82 ft) long and 3 m (9.8 ft) wide.

Prostrate leaves in Crossyne guttata.

Where leaves are basal, and lie on the ground, they are referred to as prostrate.

Basic Leaf Types

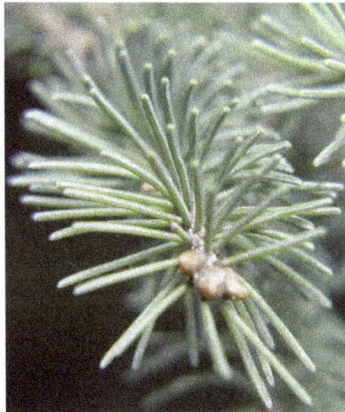

Leaves of the White Spruce (*Picea glauca*) are needle-shaped and their arrangement is spiral.

- Ferns have fronds.

- Conifer leaves are typically needle- or awl-shaped or scale-like.

- Angiosperm (flowering plant) leaves: the standard form includes stipules, a petiole, and a lamina.

- Lycophytes have microphyll leaves.

- Sheath leaves (type found in most grasses and many other monocots).

- Other specialized leaves (such as those of *Nepenthes*, a pitcher plant).

Arrangement on the Stem

Different terms are usually used to describe the arrangement of leaves on the stem (phyllotaxis):

The leaves on this plant are arranged in pairs opposite one another, with successive pairs at right angles to each other (*decussate*) along the red stem. Note the developing buds in the axils of these leaves.

- Alternate: One leaf, branch, or flower part attaches at each point or node on the stem, and leaves alternate direction, to a greater or lesser degree, along the stem.

- Basal: Arising from the base of the stem.

- Cauline: Arising from the aerial stem.

- Opposite: Two leaves, branches, or flower parts attach at each point or node on the stem. Leaf attachments are paired at each node and decussate if, as typical, each successive pair is rotated 90° progressing along the stem.

- Whorled, or verticillate: Three or more leaves, branches, or flower parts attach at each point or node on the stem. As with opposite leaves, successive whorls may or may not be decussate, rotated by half the angle between the leaves in the whorl (i.e., successive whorls of three rotated 60°, whorls of four rotated 45°, etc.). Opposite leaves may appear whorled near the tip of the stem. Pseudoverticillate describes an arrangement only appearing whorled, but not actually so.

- Rosulate: Leaves form a rosette.

- Rows: The term, *distichous*, literally means *two rows*. Leaves in this arrangement may be alternate or opposite in their attachment. The term, *2-ranked*, is equivalent. The terms, *tristichous* and *tetrastichous*, are sometimes encountered. For example, the "leaves" (actually microphylls) of most species of *Selaginella* are tetrastichous, but not decussate.

As a *stem* grows, leaves tend to appear arranged around the stem in a way that optimizes yield of light. In essence, leaves form a helix pattern centered around the stem, either clockwise or counterclockwise, with (depending upon the species) the same angle of divergence. There is a regularity in these angles and they follow the numbers in a Fibonacci sequence: 1/2, 2/3, 3/5, 5/8, 8/13, 13/21, 21/34, 34/55, 55/89. This series tends to the golden angle, which is approximately $360° \times 34/89 \approx 137.52° \approx 137° \ 30'$. In the series, the numerator indicates the number of complete turns or "gyres" until a leaf arrives at the initial position and the denominator indicates the number of leaves in the arrangement. This can be demonstrated by the following:

- Alternate leaves have an angle of 180° (or 1/2).

- 120° (or 1/3): 3 leaves in 1 circle.

- 144° (or 2/5): 5 leaves in 2 gyres.

- 135° (or 3/8): 8 leaves in 3 gyres.

Divisions of the Blade

Two basic forms of leaves can be described considering the way the blade (lamina) is divided. A simple leaf has an undivided blade. However, the leaf may be dissected to form lobes, but the gaps between lobes do not reach to the main vein. A compound leaf has a fully subdivided blade, each leaflet of the blade being separated along a main or secondary vein. The leaflets may have petiolules and stipels, the equivalents of the petioles and stipules of leaves. Because each leaflet can appear to be a simple leaf, it is important to recognize where the petiole occurs to identify a compound leaf. Compound leaves are a characteristic of some families of higher plants, such as the Fabaceae. The middle vein of a compound leaf or a frond, when it is present, is called a rachis.

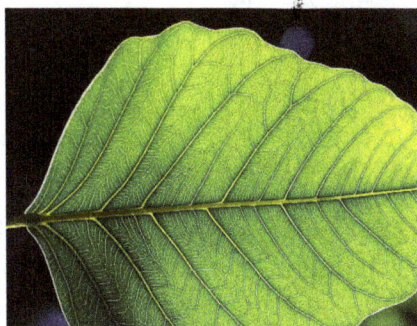

A leaf with laminar structure and pinnate venation.

- Palmately Compound: Leaves have the leaflets radiating from the end of the petiole, like fingers of the palm of a hand; for example, *Cannabis* (hemp) and *Aesculus* (buckeyes).

- Pinnately Compound: Leaves have the leaflets arranged along the main or mid-vein.

- Odd Pinnate: With a terminal leaflet; for example, *Fraxinus* (ash).

- Even Pinnate: Lacking a terminal leaflet; for example, *Swietenia* (mahogany). A specific type of even pinnate is bipinnate, where leaves only consist of two leaflets; for example, *Hymenaea*.

- Bipinnately Compound: Leaves are twice divided: the leaflets are arranged along a secondary vein that is one of several branching off the rachis. Each leaflet is called a *pinnule*. The group of pinnules on each secondary vein forms a *pinna*; for example, *Albizia* (silk tree).

- Trifoliate (or Trifoliolate): A pinnate leaf with just three leaflets; for example, *Trifolium* (clover), *Laburnum* (laburnum), and some species of *Toxicodendron* (for instance, poison ivy).

- Pinnatifid: Pinnately dissected to the central vein, but with the leaflets not entirely separate; for example, *Polypodium*, some *Sorbus* (whitebeams). In pinnately veined leaves the central vein in known as the *midrib*.

Characteristics of the Petiole

The overgrown petioles of rhubarb (*Rheum rhabarbarum*) are edible.

Petiolated leaves have a petiole (leaf stalk), and are said to be petiolate. Sessile (epetiolate) leaves have no petiole and the blade attaches directly to the stem. Subpetiolate leaves are nearly petiolate or have an extremely short petiole and may appear to be sessile.

In clasping or decurrent leaves, the blade partially surrounds the stem.

When the leaf base completely surrounds the stem, the leaves are said to be perfoliate, such as in *Eupatorium perfoliatum*. In peltate leaves, the petiole attaches to the blade inside the blade margin.

In some *Acacia* species, such as the koa tree (*Acacia koa*), the petioles are expanded or broadened and function like leaf blades; these are called phyllodes. There may or may not be normal pinnate leaves at the tip of the phyllode.

A stipule, present on the leaves of many dicotyledons, is an appendage on each side at the base of the petiole, resembling a small leaf. Stipules may be lasting and not be shed (a stipulate leaf, such as in roses and beans), or be shed as the leaf expands, leaving a stipule scar on the twig (an exstipulate leaf). The situation, arrangement, and structure of the stipules is called the "stipulation".

- Free, Lateral: As in Hibiscus.

- Adnate: Fused to the petiole base, as in Rosa.

- Ochreate: Provided with ochrea, or sheath-formed stipules, as in Polygonaceae; e.g., rhubarb.

Encircling the Petiole Base

- Interpetiolar: Between the petioles of two opposite leaves, as in Rubiaceae.

- Intrapetiolar: Between the petiole and the subtending stem, as in Malpighiaceae.

Veins

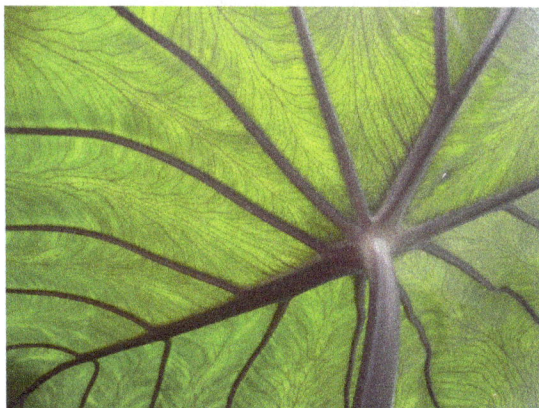

Branching veins on underside of taro leaf.

Veins (sometimes referred to as nerves) constitute one of the more visible leaf traits or characteristics. The veins in a leaf represent the vascular structure of the organ, extending into the leaf via the petiole and provide transportation of water and nutrients between leaf and stem, and play a crucial role in the maintenance of leaf water status and photosynthetic capacity. They also play a role in the mechanical support of the leaf. Within the lamina of the leaf, while some vascular plants possess only a single vein, in most this vasculature generally divides (ramifies) according to a variety of patterns (venation) and form cylindrical bundles, usually lying in the median plane of the mesophyll, between the two layers of epidermis. This pattern is often specific to taxa, and of which angiosperms possess two main types, parallel and reticulate (net like). In general, parallel venation is typical of monocots, while reticulate is more typical of eudicots and magnoliids ("dicots"), though there are many exceptions.

The venation within the bract of a lime tree.

Micrograph of a leaf skeleton.

The vein or veins entering the leaf from the petiole are called primary or first order veins. The veins branching from these are secondary or second order veins. These primary and secondary veins are considered major veins or lower order veins, though some authors include third order. Each subsequent branching is sequentially numbered, and these are the higher order veins, each branching being associated with a narrower vein diameter. In parallel veined leaves, the primary veins run parallel and equidistant to each other for most of the length of the leaf and then converge or fuse (anastomose) towards the apex. Usually many smaller minor veins interconnect these primary veins, but may terminate with very fine vein endings in the mesophyll. Minor veins are more typical of angiosperms, which may have as many as four higher orders. In contrast, leaves with reticulate venation there is a single (sometimes more) primary vein in the centre of the leaf, referred to as the midrib or costa and is continuous with the vasculature of the petiole more proximally. The midrib then branches to a number of smaller secondary veins, also known as second order veins, that extend toward the leaf margins. These often terminate in a hydathode, a secretory organ, at the margin. In turn, smaller veins branch from the secondary veins, known as tertiary or third order (or higher order) veins, forming a dense reticulate pattern. The areas or islands of mesophyll lying between the higher order veins, are called areoles. Some of the smallest veins (veinlets) may have their endings in the areoles, a process known as areolation. These minor veins act as the sites of exchange between the mesophyll and the plant's vascular system. Thus minor veins collect the products of photosynthesis (photosynthate) from the cells where it takes place, while major veins are responsible for its transport outside of the leaf. At the same time water is being transported in the opposite direction.

The number of vein endings is very variable, as is whether second order veins end at the margin, or link back to other veins. There are many elaborate variations on the patterns that the leaf veins form, and these have functional implications. Of these, angiosperms have the greatest diversity. Within these the major veins function as the support and distribution network for leaves and are correlated with leaf shape. For instance the parallel venation found in most monocots correlates with their elongated leaf shape and wide leaf base, while reticulate venation is seen in simple entire leaves, while digitate leaves typically have venation in which three or more primary veins diverge radially from a single point.

In evolutionary terms, early emerging taxa tend to have dichotomous branching with reticulate systems emerging later. Veins appeared in the Permian period (299–252 mya), prior to the appearance of angiosperms in the Triassic (252–201 mya), during which vein hierarchy appeared enabling higher function, larger leaf size and adaption to a wider vaiety of climatic conditions. Although it is the more complex pattern, branching veins appear to be plesiomorphic and in some form were present in ancient seed plants as long as 250 million years ago. A pseudo-reticulate venation that is actually a highly modified penniparallel one is an autapomorphy of some Melanthiaceae, which are monocots; e.g., *Paris quadrifolia* (True-lover's Knot). In leaves with reticulate venation, veins form a scaffolding matrix imparting mechanical rigidity to leaves.

Morphology Changes Within a Single Plant

- Homoblasty: Characteristic in which a plant has small changes in leaf size, shape, and growth habit between juvenile and adult stages, in contrast to.

- Heteroblasty: Characteristic in which a plant has marked changes in leaf size, shape, and growth habit between juvenile and adult stages.

Anatomy (Medium and Small Scale)

Medium-Scale Features

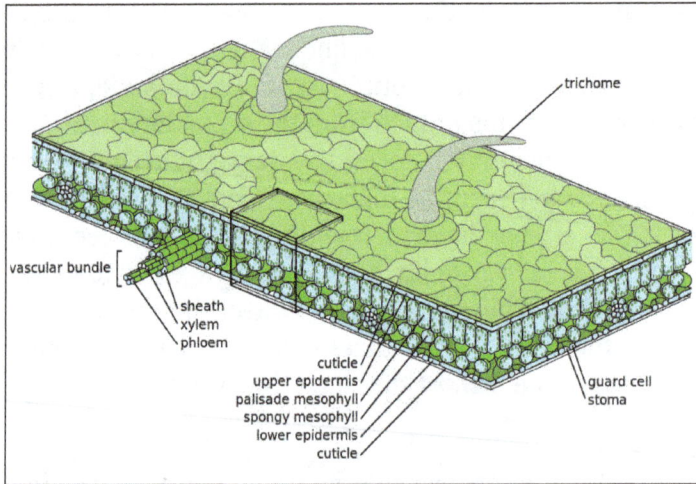

Leaves are normally extensively vascularised and typically have networks of vascular bundles containing xylem, which supplies water for photosynthesis, and phloem, which transports the sugars produced by photosynthesis. Many leaves are covered in trichomes (small hairs) which have diverse structures and functions.

Small-Scale Features

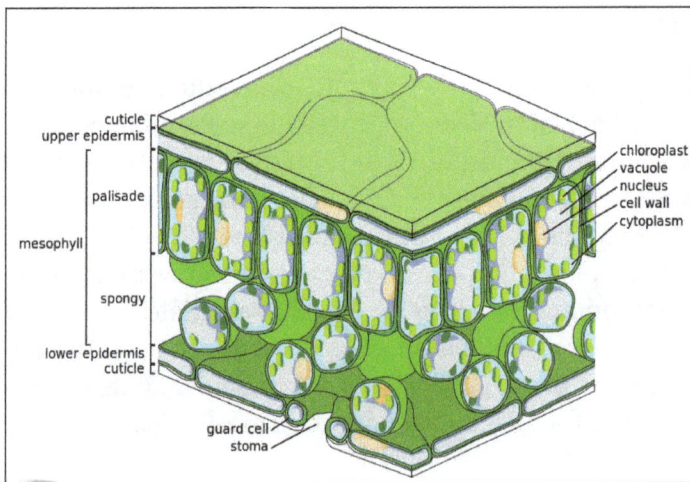

The major tissue systems present are:

- The epidermis, which covers the upper and lower surfaces.

- The mesophyll tissue inside the leaf, which is rich in chloroplasts (also called chlorenchyma).

- The arrangement of veins (the vascular tissue).

These three tissue systems typically form a regular organisation at the cellular scale. Specialised cells that differ markedly from surrounding cells, and which often synthesise specialised products such as crystals, are termed idioblasts.

Major Leaf Tissues

| Cross-section of a leaf. | Spongy mesophyll cells. | Epidermal cells. |

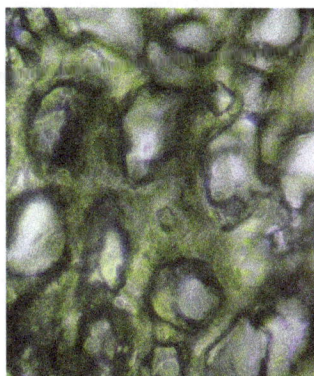

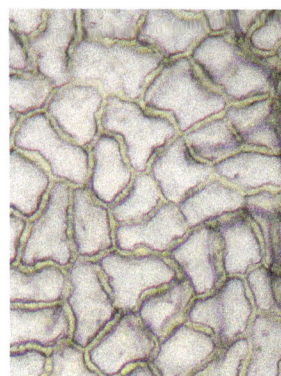

Epidermis

The epidermis is the outer layer of cells covering the leaf. It is covered with a waxy cuticle which is impermeable to liquid water and water vapor and forms the boundary separating the plant's inner cells from the external world. The cuticle is in some cases thinner on the lower epidermis than on the upper epidermis, and is generally thicker on leaves from dry climates as compared with those from wet climates. The epidermis serves several functions: protection against water loss by way of transpiration, regulation of gas exchange and secretion of metabolic compounds. Most leaves show dorsoventral anatomy: The upper (adaxial) and lower (abaxial) surfaces have somewhat different construction and may serve different functions.

The epidermis tissue includes several differentiated cell types; epidermal cells, epidermal hair cells (trichomes), cells in the stomatal complex; guard cells and subsidiary cells. The epidermal cells are the most numerous, largest, and least specialized and form the majority of the epidermis. They are typically more elongated in the leaves of monocots than in those of dicots.

Chloroplasts are generally absent in epidermal cells, the exception being the guard cells of the stomata. The stomatal pores perforate the epidermis and are surrounded on each side by chloroplast-containing guard cells, and two to four subsidiary cells that lack chloroplasts, forming a specialized cell group known as the stomatal complex. The opening and closing of the stomatal aperture is controlled by the stomatal complex and regulates the exchange of gases and water vapor between the outside air and the interior of the leaf. Stomata therefore play the important role in allowing photosynthesis without letting the leaf dry out. In a typical leaf, the stomata are more numerous over the abaxial (lower) epidermis than the adaxial (upper) epidermis and are more numerous in plants from cooler climates.

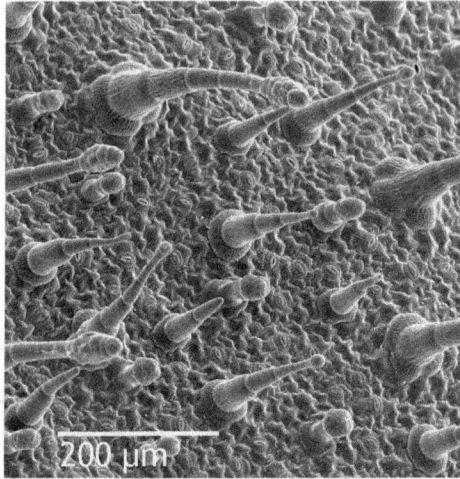

SEM image of the leaf epidermis of *Nicotiana alata*, showing trichomes (hair-like appendages) and stomata (eye-shaped slits, visible at full resolution).

Mesophyll

Most of the interior of the leaf between the upper and lower layers of epidermis is a *parenchyma* (ground tissue) or *chlorenchyma* tissue called the mesophyll (Greek for "middle leaf"). This assimilation tissue is the primary location of photosynthesis in the plant. The products of photosynthesis are called "assimilates".

In ferns and most flowering plants, the mesophyll is divided into two layers:

- An upper palisade layer of vertically elongated cells, one to two cells thick, directly beneath the adaxial epidermis, with intercellular air spaces between them. Its cells contain many more chloroplasts than the spongy layer. These long cylindrical cells are regularly arranged in one to five rows. Cylindrical cells, with the *chloroplasts* close to the walls of the cell, can take optimal advantage of light. The slight separation of the cells provides maximum absorption of carbon dioxide. Sun leaves have a multi-layered palisade layer, while shade leaves or older leaves closer to the soil are single-layered.

- Beneath the palisade layer is the spongy layer. The cells of the spongy layer are more branched and not so tightly packed, so that there are large intercellular air spaces between them for oxygen and carbon dioxide to diffuse in and out of during respiration and photosynthesis. These cells contain fewer chloroplasts than those of the palisade layer. The pores or *stomata* of the epidermis open into substomatal chambers, which are connected to the intercellular air spaces between the spongy and palisade mesophyll cells.

Leaves are normally green, due to chlorophyll in chloroplasts in the mesophyll cells. Plants that lack chlorophyll cannot photosynthesize.

Vascular Tissue

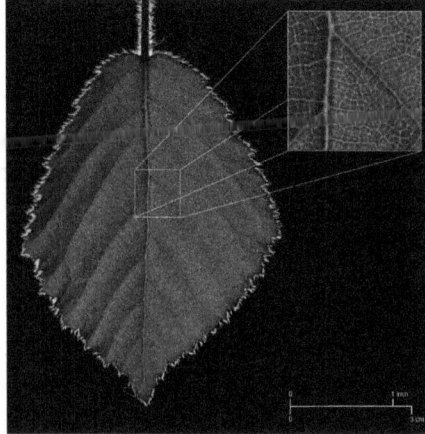

The veins of a bramble leaf.

The veins are the vascular tissue of the leaf and are located in the spongy layer of the mesophyll. The pattern of the veins is called venation. In angiosperms the venation is typically parallel in monocotyledons and forms an interconnecting network in broad-leaved plants. They were once thought to be typical examples of pattern formation through ramification, but they may instead exemplify a pattern formed in a stress tensor field.

A vein is made up of a vascular bundle. At the core of each bundle are clusters of two distinct types of conducting cells:

- Xylem: Cells that bring water and minerals from the roots into the leaf.

- Phloem: Cells that usually move sap, with dissolved sucrose(glucose to sucrose) produced by photosynthesis in the leaf, out of the leaf.

The xylem typically lies on the adaxial side of the vascular bundle and the phloem typically lies on the abaxial side. Both are embedded in a dense parenchyma tissue, called the sheath, which usually includes some structural collenchyma tissue.

Leaf Development

According to Agnes Arber's partial-shoot theory of the leaf, leaves are partial shoots, being derived from leaf primordia of the shoot apex. Compound leaves are closer to shoots than simple leaves. Developmental studies have shown that compound leaves, like shoots, may branch in three dimensions. On the basis of molecular genetics, Eckardt and Baum concluded that "it is now generally accepted that compound leaves express both leaf and shoot properties."

Ecology

Biomechanics

Plants respond and adapt to environmental factors, such as light and mechanical stress from wind. Leaves need to support their own mass and align themselves in such a way as to optimise their exposure to the sun, generally more or less horizontally. However horizontal alignment maximises exposure to bending forces and failure from stresses such as wind, snow, hail, falling debris, animals, and abrasion from surrounding foliage and plant structures. Overall leaves are relatively flimsy with regard to other plant structures such as stems, branches and roots.

Both leaf blade and petiole structure influence the leaf's response to forces such as wind, allowing a degree of repositioning to minimise drag and damage, as opposed to resistance. Leaf movement like this may also increase turbulence of the air close to the surface of the leaf, which thins the boundary layer of air immediately adjacent to the surface, increasing the capacity for gas and heat exchange, as well as photosynthesis. Strong wind forces may result in diminished leaf number and surface area, which while reducing drag, involves a trade off of also reduces photosynthesis. Thus, leaf design may involve compromise between carbon gain, thermoregulation and water loss on the one hand, and the cost of sustaining both static and dynamic loads. In vascular plants, perpendicular forces are spread over a larger area and are relatively flexible in both bending and torsion, enabling elastic deforming without damage.

Many leaves rely on hydrostatic support arranged around a skeleton of vascular tissue for their strength, which depends on maintaining leaf water status. Both the mechanics and architecture of the leaf reflect the need for transportation and support. Read and Stokes consider two basic models, the "hydrostatic" and "I-beam leaf" form. Hydrostatic leaves such as in Prostanthera lasianthos are large and thin, and may involve the need for multiple leaves rather single large leaves because of the amount of veins needed to support the periphery of large leaves. But large leaf size favours efficiency in photosynthesis and water conservation, involving further trade offs. On the other hand, I-beam leaves such as Banksia marginata involve specialised structures to stiffen them. These I-beams are formed from bundle sheath extensions of sclerenchyma meeting stiffened sub-epidermal layers. This shifts the balance from reliance on hydrostatic pressure to structural support, an obvious advantage where water is relatively scarce. Long narrow leaves bend more easily than ovate leaf blades of the same area. Monocots typically have such linear leaves that maximise surface area while minimising self-shading. In these a high proportion of longitudinal main veins provide additional support.

Interactions with other Organisms

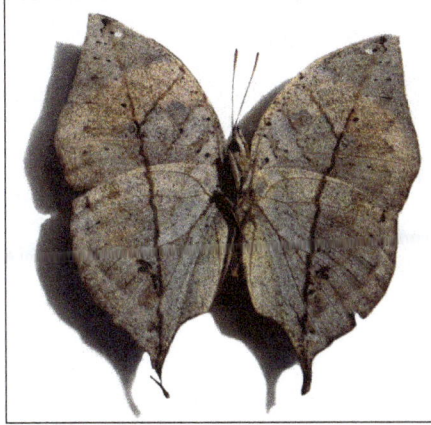

Some insects, like *Kallima inachus*, mimic leaves.

Although not as nutritious as other organs such as fruit, leaves provide a food source for many organisms. The leaf is a vital source of energy production for the plant, and plants have evolved protection against animals that consume leaves, such as tannins, chemicals which hinder the digestion of proteins and have an unpleasant taste. Animals that are specialized to eat leaves are known as folivores.

Some species have cryptic adaptations by which they use leaves in avoiding predators. For example, the caterpillars of some leaf-roller moths will create a small home in the leaf by folding it over themselves. Some sawflies similarly roll the leaves of their food plants into tubes. Females of the Attelabidae, so-called leaf-rolling weevils, lay their eggs into leaves that they then roll up as means of protection. Other herbivores and their predators mimic the appearance of the leaf. Reptiles such as some chameleons, and insects such as some katydids, also mimic the oscillating movements of leaves in the wind, moving from side to side or back and forth while evading a possible threat.

Seasonal Leaf Loss

Leaves in temperate, boreal, and seasonally dry zones may be seasonally deciduous (falling off or dying for the inclement season). This mechanism to shed leaves is called abscission. When the leaf is shed, it leaves a leaf scar on the twig. In cold autumns, they sometimes change color, and turn yellow, bright-orange, or red, as various accessory pigments (carotenoids and xanthophylls) are revealed when the tree responds to cold and reduced sunlight by curtailing chlorophyll production. Red anthocyanin pigments are now thought to be produced in the leaf as it dies, possibly to mask the yellow hue left when the chlorophyll is lost—yellow leaves appear to attract herbivores such as aphids. Optical masking of chlorophyll by anthocyanins reduces risk of photo-oxidative damage to leaf cells as they senesce, which otherwise may lower the efficiency of nutrient retrieval from senescing autumn leaves.

Leaves shifting color in autumn (fall).

Evolutionary Adaptation

In the course of evolution, leaves have adapted to different environments in the following ways:

- Waxy micro- and nanostructures on the surface reduce wetting by rain and adhesion of contamination.

- Divided and compound leaves reduce wind resistance and promote cooling.

- Hairs on the leaf surface trap humidity in dry climates and create a boundary layer reducing water loss.

- Waxy plant cuticles reduce water loss.

- Large surface area provides a large area for capture of sunlight.

- In harmful levels of sunlight, specialised leaves, opaque or partly buried, admit light through a translucent leaf window for photosynthesis at inner leaf surfaces (e.g. *Fenestraria*).

- Succulent leaves store water and organic acids for use in CAM photosynthesis.

- Aromatic oils, poisons or pheromones produced by leaf borne glands deter herbivores (e.g. eucalypts).

- Inclusions of crystalline minerals deter herbivores (e.g. silica phytoliths in grasses, raphides in Araceae).

- Petals attract pollinators.

- Spines protect the plants from herbivores (e.g. cacti).

- Stinging hairs to protect against herbivory, e.g. in *Urtica dioica* and *Dendrocnide moroides* (Urticaceae).

- Special leaves on carnivorous plants are adapted for trapping food, mainly invertebrate prey, though some species trap small vertebrates as well.

- Bulbs store food and water (e.g. onions).

- Tendrils allow the plant to climb (e.g. peas).

- Bracts and pseudanthia (false flowers) replace normal flower structures when the true flowers are greatly reduced (e.g. spurges and spathes in the Araceae.

Poinsettia bracts are leaves which have evolved red pigmentation in order to attract insects and birds to the central flowers, an adaptive function normally served by petals (which are themselves leaves highly modified by evolution).

Terminology

Shape

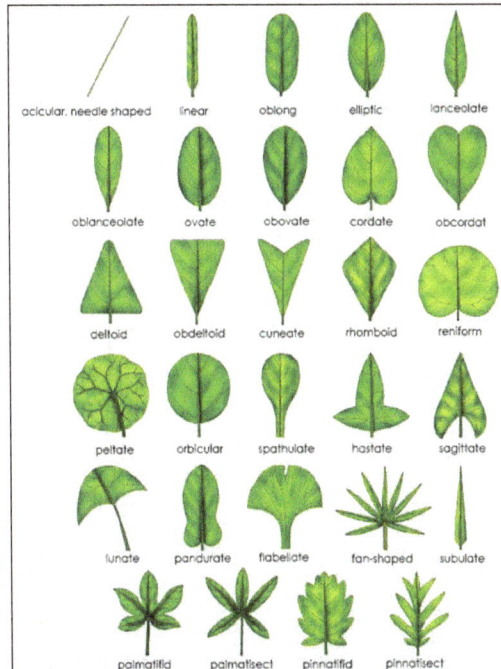

Leaf morphology terms.

Edge (Margin)

Image	Term	Description
	Entire	Even; with a smooth margin; without toothing
	Ciliate	Fringed with hairs
	Crenate	Wavy-toothed; dentate with rounded teeth
	Dentate	Toothed May be coarsely dentate, having large teeth or glandular dentate, having teeth which bear glands
	Denticulate	Finely toothed
	Doubly serrate	Each tooth bearing smaller teeth
	Serrate	Saw-toothed; with asymmetrical teeth pointing forward
	Serrulate	Finely serrate
	Sinuate	With deep, wave-like indentations; coarsely crenate
	Lobate	Indented, with the indentations not reaching the center
	Undulate	With a wavy edge, shallower than sinuate
	Spiny or pungent	With stiff, sharp points such as thistles

Apex (tip)

Image	Term	Description
	Acuminate	Long-pointed, prolonged into a narrow, tapering point in a concave manner
	Acute	Ending in a sharp, but not prolonged point

	Cuspidate	With a sharp, elongated, rigid tip; tipped with a cusp
	Emarginate	Indented, with a shallow notch at the tip
	Mucronate	Abruptly tipped with a small short point
	Mucronulate	Mucronate, but with a noticeably diminutive spine
	Obcordate	Inversely heart-shaped
	Obtuse	Rounded or blunt
	Truncate	Ending abruptly with a flat end

Base

Acuminate

Coming to a sharp, narrow, prolonged point.

Acute

Coming to a sharp, but not prolonged point.

Auriculate

Ear-shaped.

Cordate

Heart-shaped with the notch towards the stalk.

Cuneate

Wedge-shaped.

Hastate

Shaped like an halberd and with the basal lobes pointing outward.

Oblique

Slanting.

Reniform

Kidney-shaped but rounder and broader than long.

Rounded

Curving shape.

Sagittate

Shaped like an arrowhead and with the acute basal lobes pointing downward.

Truncate

Ending abruptly with a flat end, that looks cut off.

Surface

Coriaceous

Leathery; stiff and tough, but somewhat flexible.

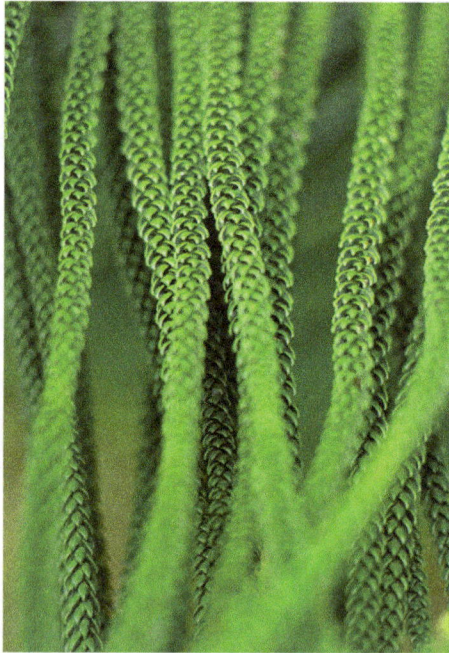

Scale-shaped leaves of a Norfolk Island Pine, *Araucaria heterophylla*.

Farinose

Bearing farina, mealy, covered with a waxy, whitish powder.

Glabrous

Smooth, not hairy.

Glaucous

With a whitish bloom; covered with a very fine, bluish-white powder.

Glutinous

Sticky, viscid.

Lepidote

Coated with small scales (thus elepidote, without such scales).

Maculate

Stained, spotted, compare immaculate.

Papillate, or Papillose

Bearing papillae (minute, nipple-shaped protuberances).

Pubescent

Covered with erect hairs (especially soft and short ones).

Punctate

Marked with dots; dotted with depressions or with translucent glands or colored dots.

Rugose

Deeply wrinkled; with veins clearly visible.

Scurfy

Covered with tiny, broad scalelike particles.

Tuberculate

Covered with tubercles; covered with warty prominences.

Verrucose

Warted, with warty outgrowths.

Viscid, or viscous

Covered with thick, sticky secretions.

The leaf surface is also host to a large variety of microorganisms; in this context it is referred to as the phyllosphere.

Hairiness

"Hairs" on plants are properly called trichomes. Leaves can show several degrees of hairiness. The meaning of several of the following terms can overlap.

Common mullein (*Verbascum thapsus*) leaves are covered in dense, stellate trichomes.

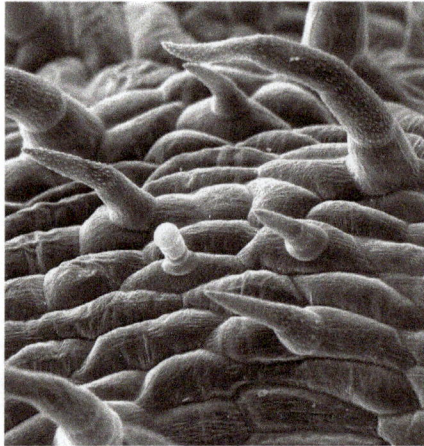

Scanning electron microscope image of trichomes on the lower surface of a *Coleus blumei* (coleus) leaf.

Arachnoid, or Arachnose

With many fine, entangled hairs giving a cobwebby appearance.

Barbellate

With finely barbed hairs (barbellae).

Bearded

With long, stiff hairs.

Bristly

With stiff hair-like prickles.

Canescent

Hoary with dense grayish-white pubescence.

Ciliate

Marginally fringed with short hairs (cilia).

Ciliolate

Minutely ciliate.

Floccose

With flocks of soft, woolly hairs, which tend to rub off.

Glabrescent

Losing hairs with age.

Glabrous

No hairs of any kind present.

Glandular

With a gland at the tip of the hair.

Hirsute

With rather rough or stiff hairs.

Hispid

With rigid, bristly hairs.

Hispidulous

Minutely hispid.

Hoary

With a fine, close grayish-white pubescence.

Lanate, or lanose

With woolly hairs.

Pilose

With soft, clearly separated hairs.

Puberulent, or puberulous

With fine, minute hairs.

Pubescent

With soft, short and erect hairs.

Scabrous, or scabrid

Rough to the touch.

Sericeous

Silky appearance through fine, straight and appressed (lying close and flat) hairs.

Silky

With adpressed, soft and straight pubescence.

Stellate, or Stelliform

With star-shaped hairs.

Strigose

With appressed, sharp, straight and stiff hairs.

Tomentose

Densely pubescent with matted, soft white woolly hairs.

Cano-tomentose

Between canescent and tomentose.

Felted-tomentose

Woolly and matted with curly hairs.

Tomentulose

Minutely or only slightly tomentose.

Villous

With long and soft hairs, usually curved.

Woolly

With long, soft and tortuous or matted hairs.

Timing

Hysteranthous

Developing after the flowers.

Synanthous

Developing at the same time as the flowers.

Venation

Classification

Hickey primary venation types:

1. Pinnate venation, Ostrya virginiana.

2. Parallel venation, *Iris*.

3. Campylodromous venation, *Maianthemum bifolium*.

4. Actinodromous venation (suprabasal), *Givotia moluccana*.

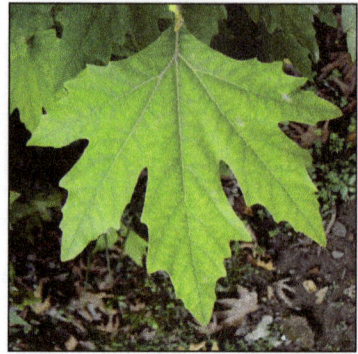

5. Acrodrous venation (basal), *Miconia calvescens.*   6. Palinactodromous venation, *Platanus orientalis.*

A number of different classification systems of the patterns of leaf veins (venation or veination) have been described, starting with Ettingshausen, together with many different descriptive terms, and the terminology has been described as "formidable". One of the commonest among these is the Hickey system, originally developed for "dicotyledons" and using a number of Ettingshausen's:

Hickey System

Pinnate (feather-veined, reticulate, pinnate-netted, penniribbed, penninerved, or pen-niveined): The veins arise pinnately (feather like) from a single primary vein (mid-vein) and subdivide into secondary veinlets, known as higher order veins. These, in turn, form a complicated network. This type of venation is typical for (but by no means limited to) "dicotyledons" (non monocotyledon angiosperms). E.g., *Ostrya.*

There are three subtypes of pinnate venation:

- *Craspedodromous:* The major veins reach to the margin of the leaf.

- *Camptodromous:* Major veins extend close to the margin, but bend before they intersect with the margin.

- *Hyphodromous:* All secondary veins are absent, rudimentary or concealed.

These in turn have a number of further subtypes such as eucamptodromous, where secondary veins curve near the margin without joining adjacent secondary veins.

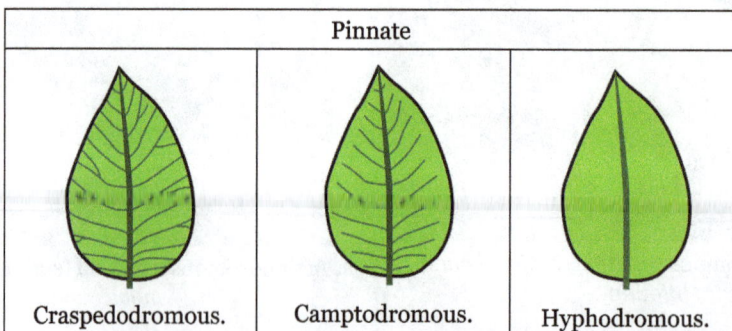

Pinnate		
Craspedodromous.	Camptodromous.	Hyphodromous.

Parallelodromous (parallel-veined, parallel-ribbed, parallel-nerved, penniparallel, striate): Two or more primary veins originating beside each other at the leaf base, and running parallel to each other to the apex and then converging there. Commissural veins (small veins) connect the major parallel veins. Typical for most monocotyledons, such as grasses.

The additional terms marginal (primary veins reach the margin), and reticulate (primary veins do not reach the margin) are also used.

Parallelodromous.

Campylodromous (*campylos* - curve): Several primary veins or branches originating at or close to a single point and running in recurved arches, then converging at apex. E.g. *Maianthemum*.

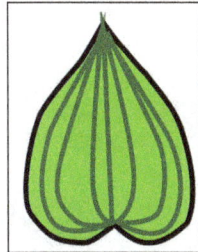

Campylodromous.

Acrodromous: Two or more primary or well developed secondary veins in convergent arches towards apex, without basal recurvature as in Campylodromous. May be basal or suprabasal depending on origin, and perfect or imperfect depending on whether they reach to 2/3 of the way to the apex. E.g., *Miconia* (basal type), *Endlicheria* (suprabasal type).

Acrodromous			
Imperfect basal.	Imperfect suprabasal.	Perfect basal.	Perfect suprabasal.

Actinodromous: Three or more primary veins diverging radially from a single point. E.g., *Arcangelisia* (basal type), *Givotia* (suprabasal type).

Actinodromous	
Imperfect marginal.	Imperfect reticulate.

Palinactodromous: Primary veins with one or more points of secondary dichotomous branching beyond the primary divergence, either closely or more distantly spaced. E.g., *Platanus*.

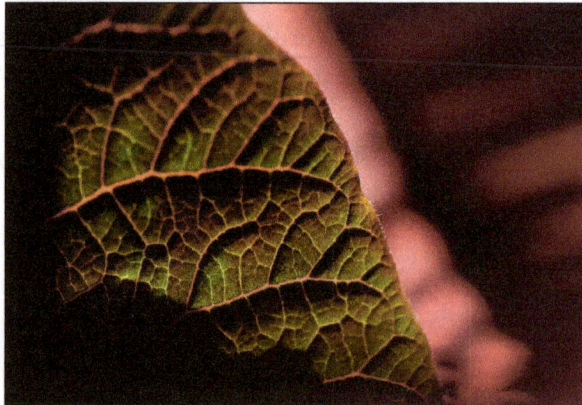

Venation of a Poinsettia (Euphorbia pulcherrima) leaf.

Palinactodromous.

Types 4–6 may similarly be subclassified as basal (primaries joined at the base of the blade) or suprabasal (diverging above the blade base), and perfect or imperfect, but also flabellate.

At about the same time, Melville (1976) described a system applicable to all Angiosperms and using Latin and English terminology. Melville also had six divisions, based on the order in which veins develop.

Arbuscular (Arbuscularis)

Branching repeatedly by regular dichotomy to give rise to a three dimensional bush-like structure consisting of linear segment (2 subclasses).

Flabellate (Flabellatus)

Primary veins straight or only slightly curved, diverging from the base in a fan-like manner (4 subclasses).

Palmate (Palmatus)

Curved primary veins (3 subclasses).

Pinnate (Pinnatus)

Single primary vein, the midrib, along which straight or arching secondary veins are arranged at more or less regular intervals (6 subclasses).

Collimate (Collimatus)

Numerous longitudinally parallel primary veins arising from a transverse meristem (5 subclasses).

Conglutinate (Conglutinatus)

Derived from fused pinnate leaflets (3 subclasses).

A modified form of the Hickey system was later incorporated into the Smithsonian classification which proposed seven main types of venation, based on the architecture of the primary veins, adding Flabellate as an additional main type. Further classification was then made on the basis of secondary veins, with 12 further types, such as;

Brochidodromous

Closed form in which the secondaries are joined together in a series of prominent arches, as in *Hildegardia*.

Craspedodromous

Open form with secondaries terminating at the margin, in toothed leaves, as in *Celtis*.

Eucamptodromous

Intermediate form with upturned secondaries that gradually diminish apically but inside the margin, and connected by intermediate tertiary veins rather than loops between secondaries, as in *Cornus*.

Cladodromous

Secondaries freely branching toward the margin, as in *Rhus*.

Terms which had been used as subtypes in the original Hickey system.

Secondary venation patterns			
Brochidodromous.	Craspedodromous.	Eucamptodromous.	Cladodromous.

Brochidodromous
Hildegardia migeodii.

Cladodromous
Rhus ovata.

Craspedodromous
Celtis occidentalis.

Eucamptodromous
Cornus officinalis.

Further descriptions included the higher order, or minor veins and the patterns of areoles.

Flabellate venation, *Adiantum cunninghamii.*

Flabellate

Several to many equal fine basal veins diverging radially at low angles and branching apically. e.g. Paranomus.

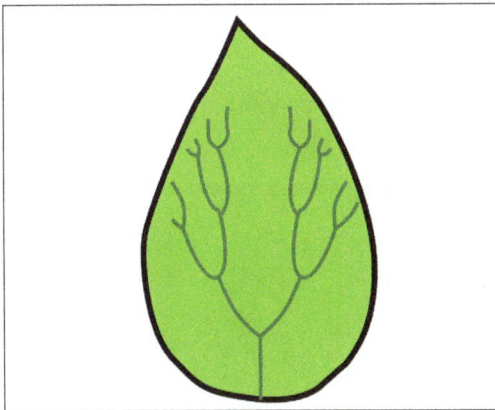

Flabellate.

Analyses of vein patterns often fall into consideration of the vein orders, primary vein type, secondary vein type (major veins), and minor vein density. A number of authors have adopted simplified versions of these schemes. At its simplest the primary vein types can be considered in three or four groups depending on the plant divisions being considered;

- Pinnate

- Palmate

- Parallel

Where palmate refers to multiple primary veins that radiate from the petiole, as opposed to branching from the central main vein in the pinnate form, and encompasses both of Hickey types 4 and 5, which are preserved as subtypes; e.g., palmate-acrodromous.

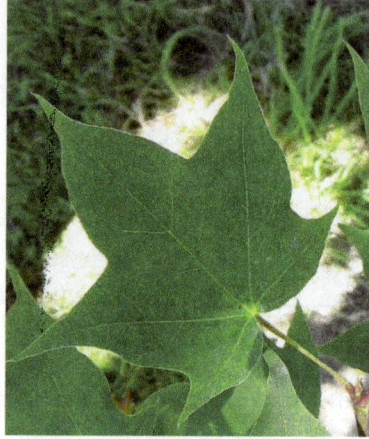

Palmate venation, *Acer truncatum.*

Palmate, Palmate-Netted, Palmate-Veined and Fan-Veined

Several main veins of approximately equal size diverge from a common point near the leaf base where the petiole attaches, and radiate toward the edge of the leaf. Palmately veined leaves are often lobed or divided with lobes radiating from the common point. They may vary in the number of primary veins (3 or more), but always radiate from a common point. e.g. most Acer (maples).

Palmate.

Other Systems

Alternatively, Simpson uses:

Uninervous

Central midrib with no lateral veins (microphyllous), seen in the non-seed bearing tracheophyt es, such as horsetails.

Dichotomous

Veins successively branching into equally sized veins from a common point, forming a Y junction, fanning out. Amongst temperate woody plants, Ginkgo biloba is the only species exhibiting dichotomous venation. Also some pteridophytes (ferns).

Parallel

Primary and secondary veins roughly parallel to each other, running the length of the leaf, often connected by short perpendicular links, rather than form networks. In some species, the parallel veins join together at the base and apex, such as needle-type evergreens and grasses. Characteristic of monocotyledons, but exceptions include Arisaema, and as below, under netted.

Netted (Reticulate, Pinnate)

A prominent midvein with secondary veins branching off along both sides of it. The name derives from the ultimate veinlets which form an interconnecting net like pattern or network. (The primary and secondary venation may be referred to as pinnate, while the net like finer veins are referred to as netted or reticulate); most non-monocot angiosperms, exceptions including Calophyllum. Some monocots have reticulate venation, including Colocasia, Dioscorea and Smilax.

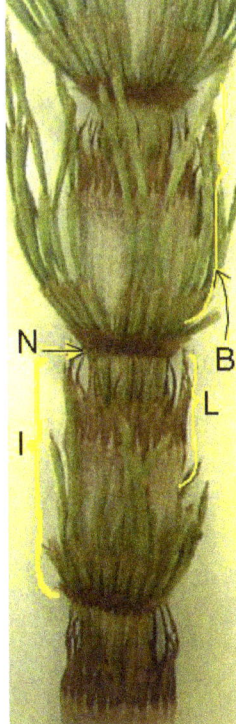

Equisetum: Reduced microphyllous leaves (L) arising in whorl from node.

However, these simplified systems allow for further division into multiple subtypes.

Simpson, (and others) divides parallel and netted (and some use only these two terms for Angiosperms) on the basis of the number of primary veins (costa) as follows;

Ginkgobiloba: Dichotomous venation.

Parallel

Penni-parallel (pinnate, pinnate parallel, unicostate parallel): Single central prominent midrib, secondary veins from this arise perpendicularly to it and run parallel to each other towards the margin or tip, but do not join (anastomose). The term unicostate refers to the prominence of the single midrib (costa) running the length of the leaf from base to apex. e.g. Zingiberales, such as Bananas etc.

Palmate-parallel (multicostate parallel): Several equally prominent primary veins arising from a single point at the base and running parallel towards tip or margin. The term multicostate refers to having more than one prominent main vein. e.g. "fan" (palmate) palms (Arecaceae).

Multicostate parallel convergent: Mid-veins converge at apex e.g. *Bambusa arundinacea = B. bambos* (Aracaceae), *Eichornia*.

Multicostate parallel divergent: Mid-veins diverge more or less parallel towards the margin e.g. *Borassus* (Poaceae), fan palms.

Netted (Reticulate)

Pinnately (veined, netted, unicostate reticulate): Single prominent midrib running from base to apex, secondary veins arising on both sides along the length of the primary midrib, running towards the margin or apex (tip), with a network of smaller veinlets forming a reticulum (mesh or network). e.g. *Mangifera, Ficus religiosa, Psidium guajava, Hibiscus rosa-sinensis, Salix alba*.

Palmately (multicostate reticulate): More than one primary veins arising from a single point, running from base to apex. e.g. *Liquidambar styraciflua* This may be further subdivided.

Multicostate convergent: Major veins diverge from origin at base then converge towards the tip. e.g. *Zizyphus, Smilax, Cinnamomum*.

Multicostate divergent: All major veins diverge towards the tip. e.g. *Gossypium, Cucurbita, Carica papaya, Ricinus communis.*

Ternately (ternate-netted): Three primary veins, as above, e.g. *Ceanothus leucodermis, C. tomentosus, Encelia farinosa.*

Simpson venation patterns

Maranta leuconeura var. *erythroneura* (Zingiberales): Penni-parallel.

Coccothrinax argentea (Arecaceae): Palmate-parallel.

Ziziphus jujuba: Multicostate palmate convergent.

Liquidambar styraciflua: Palmately netted.

Borassus sp.: Multicostate parallel divergent.

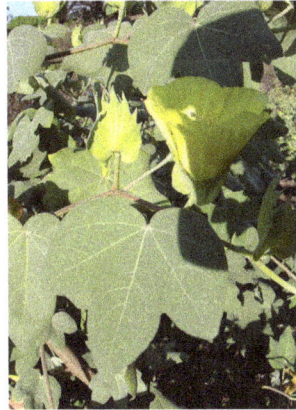

Gossypium tomentosum: Multicostate palmate divergent.

These complex systems are not used much in morphological descriptions of taxa, but have usefulness in plant identification, although criticized as being unduly burdened with jargon.

An older, even simpler system, used in some flora uses only two categories, open and closed.

- Open: Higher order veins have free endings among the cells and are more characteristic of non-monocotyledon angiosperms. They are more likely to be associated with leaf shapes that are toothed, lobed or compound. They may be subdivided as;

  - Pinnate (feather-veined) leaves, with a main central vein or rib (midrib), from which the remainder of the vein system arises.

  - Palmate, in which three or more main ribs rise together at the base of the leaf, and diverge upward.

  - Dichotomous, as in ferns, where the veins fork repeatedly.

- Closed: Higher order veins are connected in loops without ending freely among the cells. These tend to be in leaves with smooth outlines, and are characteristic of monocotyledons.

  - They may be subdivided into whether the veins run parallel, as in grasses, or have other patterns.

Other Descriptive Terms

There are also many other descriptive terms, often with very specialised usage and confined to specific taxonomic groups. The conspicuousness of veins depends on a number of features. These include the width of the veins, their prominence in relation to the lamina surface and the degree of opacity of the surface, which may hide finer veins. In this regard, veins are called obscure and the order of veins that are obscured and whether upper, lower or both surfaces, further specified.

Terms that describe vein prominence include bullate, channelled, flat, guttered, impressed, prominent and recessed. Veins may show different types of prominence in different areas of the leaf. For instance *Pimenta racemosa* has a channelled midrib on the upper surfae, but this is prominent on the lower surface. Describing vein prominence:

Bullate

Surface of leaf raised in a series of domes between the veins on the upper surface, and therefore also with marked depressions. e.g. Rytigynia pauciflora, Vitis vinifera.

Channelled (Canaliculate)

Veins sunken below the surface, resulting in a rounded channel. Sometimes confused with "guttered" because the channels may function as gutters for rain to run off and allow drying, as in many Melastomataceae. e.g. Pimenta racemosa (Myrtaceae), Clidemia hirta (Melastomataceae).

Guttered

Veins partly prominent, the crest above the leaf lamina surface, but with channels running along each side, like gutters.

Impressed

Vein forming raised line or ridge which lies below the plane of the surface which bears it, as if pressed into it, and are often exposed on the lower surface. Tissue near the veins often appears to pucker, giving them a sunken or embossed appearance.

Obscure

Veins not visible, or not at all clear; if unspecified, then not visible with the naked eye. e.g. Berberis gagnepainii. In this Berberis, the veins are only obscure on the undersurface.

Prominent

Vein raised above surrounding surface so to be easily felt when stroked with finger. e.g. Pimenta racemosa, Spathiphyllum cannifolium.

Recessed

Vein is sunk below the surface, more prominent than surrounding tissues but more sunken in channel than with impressed veins. e.g. Viburnum plicatum.

Types of vein prominence

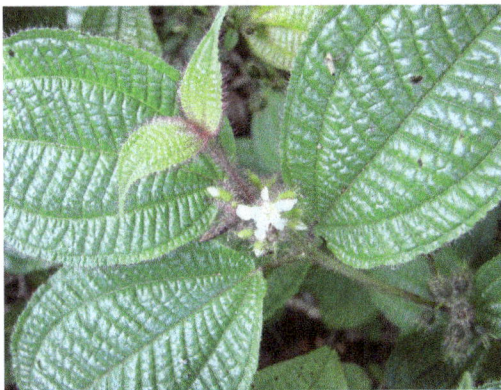

| *Clidemia hirta* Channeled. | *Berberis gagnepainii* Obscure (under surface). |

Viburnum plicatum Recessed.

Spathiphyllum cannifolium Prominent.

Cornus mas: Impressed.

Describing other features:

Plinervy (Plinerved)

More than one main vein (nerve) at the base. Lateral secondary veins branching from a point above the base of the leaf. Usually expressed as a suffix, as in 3-plinerved or triplinerved leaf. In a 3-plinerved (triplinerved) leaf three main veins branch above the base of the lamina (two secondary veins and the main vein) and run essentially parallel subsequently, as in Ceanothus and in Celtis. Similarly a quintuplinerve (five-veined) leaf has four secondary veins and a main vein. A pattern with 3-7 veins is especially conspicuous in Melastomataceae. The term has also been used in Vaccinieae. The term has been used as synonymous with acrodromous, palmate-acrodromous or suprabasal acrodromous, and is thought to be too broadly defined.

Scalariform

Veins arranged like the rungs of a ladder, particularly higher order veins.

Submarginal

Veins running close to leaf margin.

Trinerved

2 major basal nerves besides the midrib.

Diagrams of Venation Patterns

Image	Term	Description
	Arcuate	Secondary arching toward the apex
	Dichotomous	Veins splitting in two
	Longitudinal	All veins aligned mostly with the midvein
	Parallel	All veins parallel and not intersecting
	Pinnate	Secondary veins borne from midrib
	Reticulate	All veins branching repeatedly, net veined
	Rotate	Veins coming from the center of the leaf and radiating toward the edges
	Transverse	Tertiary veins running perpendicular to axis of main vein, connecting secondary veins

Size

The terms megaphyll, macrophyll, mesophyll, notophyll, microphyll, nanophyll and leptophyll are used to describe leaf sizes (in descending order), in a classification devised in 1934 by Christen C. Raunkiær and since modified by others.

The Stem

Stem is the plant axis that bears buds and shoots with leaves and, at its basal end, roots. The stem conducts water, minerals, and food to other parts of the plant; it may also store food, and green stems themselves produce food. In most plants the stem is the major vertical shoot, in some it is inconspicuous, and in others it is modified and resembles other plant parts (e.g., underground stems may look like roots).

The primary functions of the stem are to support the leaves; to conduct water and minerals to the leaves, where they can be converted into usable products by photosynthesis; and to transport these products from the leaves to other parts of the plant, including the roots. The stem conducts water and nutrient minerals from their site of absorption in the roots to the leaves by means of certain vascular tissues in the xylem. The movement of synthesized foods from the leaves to other plant organs occurs chiefly through other vascular tissues in the stem called phloem. Food and water are also frequently stored in the stem. Examples of food-storing stems include such specialized forms as tubers, rhizomes, and corms and the woody stems of trees and shrubs. Water storage is developed to a high degree in the stems of cacti, and all green stems are capable of photosynthesis.

Growth and Anatomy

The first rudiment of the young stem, or shoot, of an embryonic plant appears from the seed after the root has first protruded. The growing portion at the apex of the shoot is the terminal bud of the plant, and by the continued development of this bud and its adjacent tissues, the stem increases in height. Lateral buds and leaves grow out of the stem at intervals called nodes; the intervals on the stem between the nodes are called internodes. The number of leaves that appear at a node depends on the species of plant; one leaf per node is common, but two or or more leaves may grow at the nodes of some species. When a leaf drops off a stem at the end of a growing season, it leaves a scar on the stem because of the severing of the vascular (conducting) bundles that had connected stem and leaf. As the stem continues to grow, lateral buds are produced that develop into lateral shoots more or less resembling the parent stem, and these ultimately determine the branching of the plant. In trees the lateral shoots develop into branches, from which other lateral shoots, called branchlets, or twigs, arise. The point at which a leaf diverges in axis from a stem is called the axil. A bud formed in the axil of a previously formed leaf is called an axillary bud, and it, like the leaves, is produced from the tissues of the stem. During the development of such buds, vascular bundles are formed within them that are continuous with those of the stem.

In the stems of young dicotyledons (angiosperms with two seed leaves) and gymnosperms, the vascular bundles (xylem and phloem) are arranged in a circle around a central core of spongy ground tissue called the pith. Surrounding the vascular bundles is a layer that varies in thickness in different species and is called the cortex. Surrounding this and comprising the exterior surface of the stem is a layer called the epidermis. In plants with woody stems, a variety of secondary tissues are added to these primary tissues. Among the most important of these is a ring of meristematic cells that in turn give rise to the vascular cambium. This tissue arises between the primary xylem and phloem and gives rise to secondary phloem on the outside and secondary xylem on the inside; the latter tissue is the wood of trees.

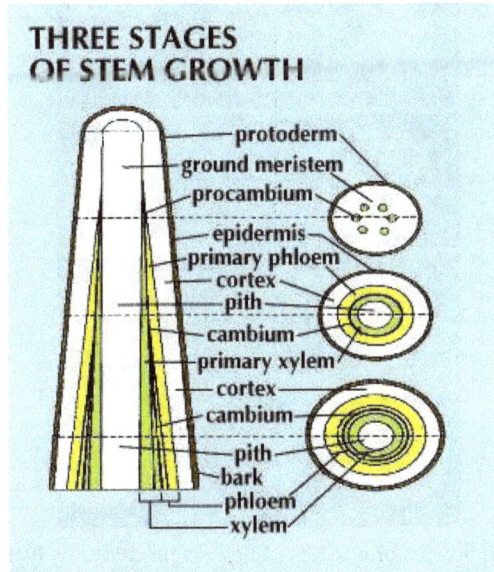

THREE STAGES
OF STEM GROWTH

protoderm
ground meristem
procambium
epidermis
primary phloem
cortex
pith
cambium
primary xylem
cortex
cambium
pith
bark
phloem
xylem

Stem anatomy: A longitudinal section, left, and cross section, right, of a growing stem show the organization of various tissues for younger, top, and older, bottom, parts of the stem.

Stem Types and Modifications

Many plants are annuals and complete their life cycles in one growing season, after which the entire plant, including the stem, dies. In biennial plants the lower part of the stem, often modified for food storage, persists after the first growing season and bears buds from which an erect stem arises during the second growing season. In perennial plants the short stem may produce new shoots for many years. Plants producing woody stems are called trees and shrubs; the latter produce branches from or near the ground, while the former have conspicuous trunks.

In general, the habit of a stem is erect or ascending, but it may lie prostrate on the ground, as in the sweet potato and strawberry. A stem may climb on rocks or plants by means of rootlets, as in ivy; other vines have twining stems that twist around a supporting plant in a spiral manner, as in the honeysuckle and hop. In other cases, climbing plants are supported by tendrils that may be specialized stems, as in the grape and passion-flower. In tropical climates twining plants often form thick woody stems and are called lianas, while in temperate regions they are generally herbaceous vines. A stolon is a stem that curves toward the ground and, on reaching a moist spot, takes root and forms an upright stem and ultimately a separate plant. Among the subterranean stems are the rhizome, corm, and tuber. In some plants the stem does not elongate during its early development but instead forms a short conical structure from which a crown of leaves arises. These may form a bulb (as in the onion and lily), a head (cabbage, lettuce), or a rosette (dandelion, plantain).

The Flower

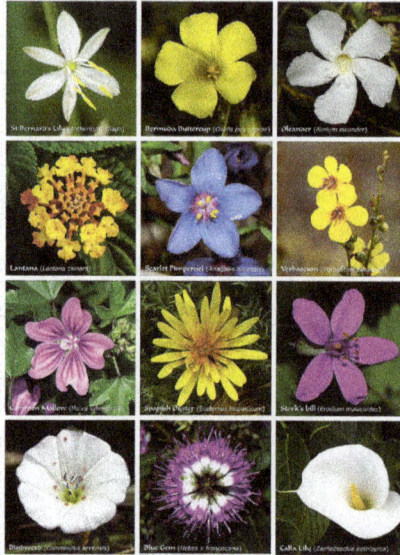

A poster with flowers or clusters of flowers produced by twelve species of flowering plants from different families.

Flowers in the Netherlands.

A flower, sometimes known as a bloom or blossom, is the reproductive structure found in flowering plants (plants of the division Magnoliophyta, also called angiosperms). The biological function of a flower is to effect reproduction, usually by providing a mechanism for the union of sperm with eggs. Flowers may facilitate outcrossing (fusion of sperm and eggs from different individuals in a population) or allow selfing (fusion of sperm and egg from the same flower). Some flowers produce diaspores without fertilization (parthenocarpy). Flowers contain sporangia and are the site where gametophytes develop. Many flowers have evolved to be attractive to animals, so as to cause them to be vectors for the transfer of pollen. After fertilization, the ovary of the flower develops into fruit containing seeds.

In addition to facilitating the reproduction of flowering plants, flowers have long been admired and used by humans to bring beauty to their environment, and also as objects of romance, ritual, religion, medicine and as a source of food.

Morphology

Floral Parts

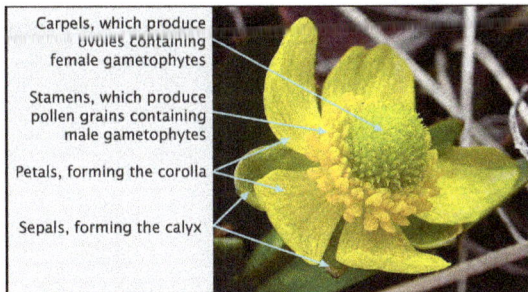

Main parts of a mature flower (Ranunculus glaberrimus).

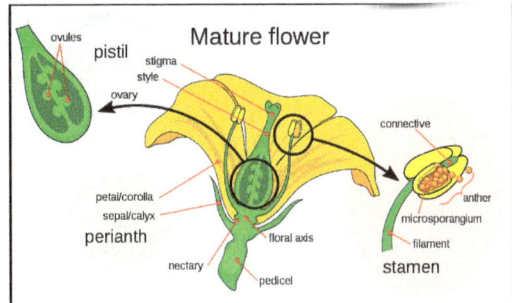

Diagram of flower parts.

The essential parts of a flower can be considered in two parts: the vegetative part, consisting of petals and associated structures in the perianth, and the reproductive or sexual parts. A stereotypical flower consists of four kinds of structures attached to the tip of a short stalk. Each of these kinds of parts is arranged in a whorl on the receptacle. The four main whorls (starting from the base of the flower or lowest node and working upwards) are as follows:

Perianth

Collectively the calyx and corolla form the perianth:

- *Calyx*: the outermost whorl consisting of units called *sepals*; these are typically green and enclose the rest of the flower in the bud stage, however, they can be absent or prominent and petal-like in some species.

- *Corolla*: the next whorl toward the apex, composed of units called *petals*, which are typically thin, soft and colored to attract animals that help the process of pollination.

Reproductive

- *Androecium*: the next whorl (sometimes multiplied into several whorls), consisting of units called stamens. Stamens consist of two parts: a stalk called a filament, topped by an anther where pollen is produced by meiosis and eventually dispersed.

- *Gynoecium*: the innermost whorl of a flower, consisting of one or more units called carpels. The carpel or multiple fused carpels form a hollow structure called an ovary, which produces ovules internally. Ovules are megasporangia and they in turn produce megaspores by meiosis which develop into female gametophytes. These give rise to egg cells. The gynoecium of a flower is also described using an alternative terminology wherein the structure one sees in the innermost whorl (consisting of an ovary, style and stigma) is called a pistil. A pistil may consist of a single carpel or a number of carpels fused together. The sticky tip of the pistil, the stigma, is the receptor of pollen. The supportive stalk, the style, becomes the pathway for pollen tubes to grow from pollen grains adhering to the stigma. The relationship to the gynoecium on the receptacle is described as hypogynous (beneath a superior ovary), perigynous (surrounding a superior ovary), or epigynous (above inferior ovary).

Reproductive parts of Easter Lily (*Lilium longiflorum*).
1. Stigma, 2. Style, 3. Stamens, 4. Filament, 5. Petal.

Structure

Although the arrangement described above is considered "typical", plant species show a wide variation in floral structure. These modifications have significance in the evolution of flowering plants and are used extensively by botanists to establish relationships among plant species.

The four main parts of a flower are generally defined by their positions on the receptacle and not by their function. Many flowers lack some parts or parts may be modified into other functions and/or look like what is typically another part. In some families, like Ranunculaceae, the petals are greatly reduced and in many species the sepals are colorful and petal-like. Other flowers have modified stamens that are petal-like; the double flowers of Peonies and Roses are mostly petaloid stamens. Flowers show great variation and plant scientists describe this variation in a systematic way to identify and distinguish species.

Specific terminology is used to describe flowers and their parts. Many flower parts are fused together; fused parts originating from the same whorl are connate, while fused parts originating from different whorls are adnate; parts that are not fused are free. When petals are fused into a tube or ring that falls away as a single unit, they are sympetalous (also called gamopetalous). Connate petals may have distinctive regions: the cylindrical base is the tube, the expanding region is the throat and the flaring outer region is the limb. A sympetalous flower, with bilateral symmetry with an upper and lower lip, is bilabiate. Flowers with connate petals or sepals may have various shaped corolla or calyx, including campanulate, funnelform, tubular, urceolate, salverform or rotate.

Referring to "fusion," as it is commonly done, appears questionable because at least some of the processes involved may be non-fusion processes. For example, the addition of intercalary growth at or below the base of the primordia of floral appendages such as sepals, petals, stamens and carpels may lead to a common base that is not the result of fusion.

Many flowers have a symmetry. When the perianth is bisected through the central axis from any point and symmetrical halves are produced, the flower is said to be actinomorphic or regular, e.g. rose or trillium. This is an example of radial symmetry. When flowers are bisected and produce only one line that produces symmetrical halves, the flower is said to be irregular or zygomorphic, e.g. snapdragon or most orchids.

Left: A normal zygomorphic Streptocarpus flower. Right: An aberrant peloric Streptocarpus flower. Both of these flowers appeared on the Streptocarpus hybrid 'Anderson's Crows' Wings'.

Flowers may be directly attached to the plant at their base (sessile—the supporting stalk or stem is highly reduced or absent). The stem or stalk subtending a flower is called a peduncle. If a peduncle supports more than one flower, the stems connecting each flower to the main axis are called pedicels. The apex of a flowering stem forms a terminal swelling which is called the *torus* or receptacle.

Inflorescence

In those species that have more than one flower on an axis, the collective cluster of flowers is termed an *inflorescence*. Some inflorescences are composed of many small flowers arranged in a formation that resembles a single flower. The common example of this is most members of the very large composite (Asteraceae) group. A single daisy or sunflower, for example, is not a flower but a flower *head*—an inflorescence composed of numerous flowers (or florets). An inflorescence may include specialized stems and modified leaves known as bracts.

The familiar calla lily is not a single flower. It is actually an inflorescence of tiny flowers pressed together on a central stalk that is surrounded by a large petal-like bract.

Floral Diagrams and Floral Formulae

A *floral formula* is a way to represent the structure of a flower using specific letters, numbers and symbols, presenting substantial information about the flower in a compact form. It can represent a taxon, usually giving ranges of the numbers of different organs, or particular species. Floral formulae have been developed in the early 19th century and their use has declined since. Prenner *et al.* (2010) devised an extension of the existing model to broaden the descriptive capability of the formula. The format of floral formulae differs in different parts of the world, yet they convey the same information.

The structure of a flower can also be expressed by the means of *floral diagrams*. The use of schematic diagrams can replace long descriptions or complicated drawings as a tool for understanding both floral structure and evolution. Such diagrams may show important features of flowers, including the relative positions of the various organs, including the presence of fusion and symmetry, as well as structural details.

Development

A flower develops on a modified shoot or axis from a determinate apical meristem (*determinate* meaning the axis grows to a set size). It has compressed internodes, bearing structures that in classical plant morphology are interpreted as highly modified leaves. Detailed developmental studies, however, have shown that stamens are often initiated more or less like modified stems (caulomes) that in some cases may even resemble branchlets. Taking into account the whole diversity in the development of the androecium of flowering plants, we find a continuum between modified leaves (phyllomes), modified stems (caulomes), and modified branchlets (shoots).

Flowering Transition

The transition to flowering is one of the major phase changes that a plant makes during its life cycle. The transition must take place at a time that is favorable for fertilization and the formation of seeds, hence ensuring maximal reproductive success. To meet these needs a plant is able to interpret important endogenous and environmental cues such as changes in levels of plant hormones and seasonable temperature and photoperiod changes. Many perennial and most biennial plants require vernalization to flower. The molecular interpretation of these signals is through the transmission of a complex signal known as florigen, which involves a variety of genes, including Constans, Flowering Locus C and Flowering Locus T. Florigen is produced in the leaves in reproductively favorable conditions and acts in buds and growing tips to induce a number of different physiological and morphological changes.

The first step of the transition is the transformation of the vegetative stem primordia into floral primordia. This occurs as biochemical changes take place to change cellular differentiation of leaf, bud and stem tissues into tissue that will grow into the reproductive organs. Growth of the central part of the stem tip stops or flattens out and the sides develop protuberances in a whorled or spiral fashion around the outside of the stem end. These protuberances develop into the sepals, petals, stamens, and carpels. Once this process begins, in most plants, it cannot be reversed and the stems develop flowers, even if the initial start of the flower formation event was dependent of some environmental cue. Once the process begins, even if that cue is removed the stem will continue to develop a flower.

Yvonne Aitken has shown that flowering transition depends on a number of factors, and that plants flowering earliest under given conditions had the least dependence on climate whereas later-flowering varieties reacted strongly to the climate setup.

Organ Development

The molecular control of floral organ identity determination appears to be fairly well understood in some species. In a simple model, three gene activities interact in a combinatorial manner to determine the developmental identities of the organ primordia within the floral meristem. These gene functions are called A, B and C-gene functions. In the first floral whorl only A-genes are expressed, leading to the formation of sepals. In the second whorl both A- and B-genes are expressed, leading to the formation of petals. In the third whorl, B and C genes interact to form stamens and in the center of the flower C-genes alone give rise to carpels. The model is based upon studies of mutants in

Arabidopsis thaliana and snapdragon, *Antirrhinum majus*. For example, when there is a loss of B-gene function, mutant flowers are produced with sepals in the first whorl as usual, but also in the second whorl instead of the normal petal formation. In the third whorl the lack of B function but presence of C-function mimics the fourth whorl, leading to the formation of carpels also in the third whorl.

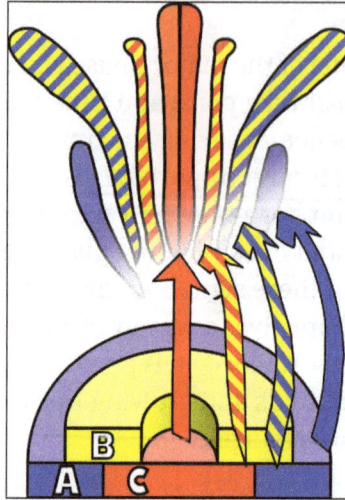

The ABC model of flower development.

Most genes central in this model belong to the MADS-box genes and are transcription factors that regulate the expression of the genes specific for each floral organ.

Floral Function

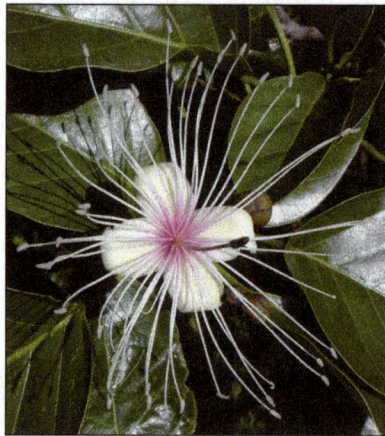

A "perfect flower", this *Crateva religiosa* flower has both stamens (outer ring) and a pistil (center).

The principal purpose of a flower is the reproduction of the individual and the species. All flowering plants are *heterosporous*, producing two types of spores. Microspores are produced by meiosis inside anthers while megaspores are produced inside ovules, inside an ovary. In fact, anthers typically consist of four microsporangia and an ovule is an integumented megasporangium. Both types of spores develop into gametophytes inside sporangia. As with all heterosporous plants, the gametophytes also develop inside the spores (are endosporic).

In the majority of species, individual flowers have both functional carpels and stamens. Botanists describe these flowers as being *perfect* or *bisexual* and the species as *hermaphroditic*. Some flowers lack one or the other reproductive organ and called *imperfect* or *unisexual*. If unisex flowers are found on the same individual plant but in different locations, the species is said to be *monoecious*. If each type of unisex flower is found only on separate individuals, the plant is *dioecious*.

Flower Specialization and Pollination

Flowering plants usually face selective pressure to optimize the transfer of their pollen, and this is typically reflected in the morphology of the flowers and the behaviour of the plants. Pollen may be transferred between plants via a number of 'vectors'. Some plants make use of abiotic vectors — namely wind (anemophily) or, much less commonly, water (hydrophily). Others use biotic vectors including insects (entomophily), birds (ornithophily), bats (chiropterophily) or other animals. Some plants make use of multiple vectors, but many are highly specialised.

Cleistogamous flowers are self-pollinated, after which they may or may not open. Many Viola and some Salvia species are known to have these types of flowers.

The flowers of plants that make use of biotic pollen vectors commonly have glands called nectaries that act as an incentive for animals to visit the flower. Some flowers have patterns, called nectar guides, that show pollinators where to look for nectar. Flowers also attract pollinators by scent and color. Still other flowers use mimicry to attract pollinators. Some species of orchids, for example, produce flowers resembling female bees in color, shape, and scent. Flowers are also specialized in shape and have an arrangement of the stamens that ensures that pollen grains are transferred to the bodies of the pollinator when it lands in search of its attractant (such as nectar, pollen, or a mate). In pursuing this attractant from many flowers of the same species, the pollinator transfers pollen to the stigmas—arranged with equally pointed precision—of all of the flowers it visits.

Anemophilous flowers use the wind to move pollen from one flower to the next. Examples include grasses, birch trees, ragweed and maples. They have no need to attract pollinators and therefore tend not to be "showy" flowers. Male and female reproductive organs are generally found in separate flowers, the male flowers having a number of long filaments terminating in exposed stamens, and the female flowers having long, feather-like stigmas. Whereas the pollen of animal-pollinated flowers tends to be large-grained, sticky, and rich in protein (another "reward" for pollinators), anemophilous flower pollen is usually small-grained, very light, and of little nutritional value to animals.

Pollination

Grains of pollen sticking to this bee will be transferred to the next flower it visits.

The primary purpose of a flower is reproduction. Since the flowers are the reproductive organs of plant, they mediate the joining of the sperm, contained within pollen, to the ovules — contained in the ovary. Pollination is the movement of pollen from the anthers to the stigma. The joining of the sperm to the ovules is called fertilization. Normally pollen is moved from one plant to another, but many plants are able to self pollinate. The fertilized ovules produce seeds that are the next generation. Sexual reproduction produces genetically unique offspring, allowing for adaptation. Flowers have specific designs which encourages the transfer of pollen from one plant to another of the same species. Many plants are dependent upon external factors for pollination, including: wind and animals, and especially insects. Even large animals such as birds, bats, and pygmy possums can be employed. The period of time during which this process can take place (the flower is fully expanded and functional) is called *anthesis*. The study of pollination by insects is called *anthecology*.

Pollination Mechanism

The pollination mechanism employed by a plant depends on what method of pollination is utilized.

Most flowers can be divided between two broad groups of pollination methods:

- *Entomophilous*: flowers attract and use insects, bats, birds or other animals to transfer pollen from one flower to the next. Often they are specialized in shape and have an arrangement of the stamens that ensures that pollen grains are transferred to the bodies of the pollinator when it lands in search of its attractant (such as nectar, pollen, or a mate). In pursuing this attractant from many flowers of the same species, the pollinator transfers pollen to the stigmas—arranged with equally pointed precision—of all of the flowers it visits. Many flowers rely on simple proximity between flower parts to ensure pollination. Others, such as the *Sarracenia* or lady-slipper orchids, have elaborate designs to ensure pollination while preventing self-pollination.

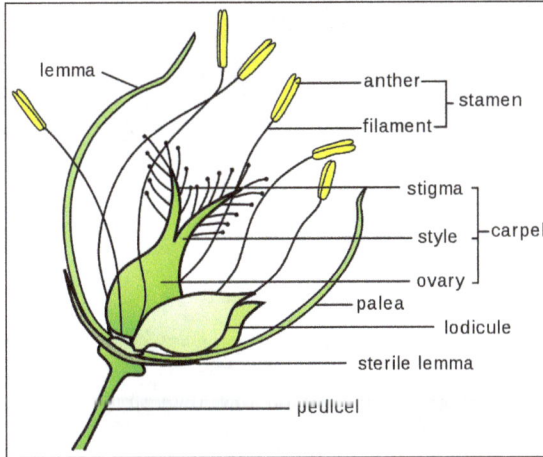

Grass flower with vestigial perianth or lodicules.

- *Anemophilous*: flowers use the wind to move pollen from one flower to the next, examples include the grasses, Birch trees, Ragweed and Maples. They have no need to attract pollinators and therefore tend not to grow large blossoms. Whereas the pollen of entomophilous flowers tends to be large-grained, sticky, and rich in protein (another "reward" for pollinators), anemophilous flower pollen is usually small-grained, very light, and of little nutritional value to insects, though it may still be gathered in times of dearth. Honeybees and bumblebees actively gather anemophilous corn (maize) pollen, though it is of little value to them.

Some flowers with both stamens and a pistil are capable of self-fertilization, which does increase the chance of producing seeds but limits genetic variation. The extreme case of self-fertilization occurs in flowers that always self-fertilize, such as many dandelions. Some flowers are self-pollinated and use flowers that never open or are self-pollinated before the flowers open, these flowers are called cleistogamous. Many Viola species and some Salvia have these types of flowers. Conversely, many species of plants have ways of preventing self-fertilization. Unisexual male and female flowers on the same plant may not appear or mature at the same time, or pollen from the same plant may be incapable of fertilizing its ovules. The latter flower types, which have chemical barriers to their own pollen, are referred to as self-sterile or self-incompatible.

Attraction Methods

Plants cannot move from one location to another, thus many flowers have evolved to attract animals to transfer pollen between individuals in dispersed populations. Flowers that are insect-pollinated are called *entomophilous*; literally "insect-loving" in Greek. They can be highly modified along with the pollinating insects by co-evolution. Flowers commonly have glands called *nectaries* on various parts that attract animals looking for nutritious nectar. Birds and bees have color vision, enabling them to seek out "colorful" flowers.

A Bee orchid has evolved over many generations to better mimic a female bee to attract male bees as pollinators.

Some flowers have patterns, called nectar guides, that show pollinators where to look for nectar; they may be visible only under ultraviolet light, which is visible to bees and some other insects. Flowers also attract pollinators by scent and some of those scents are pleasant to our sense of smell. Not all flower scents are appealing to humans; a number of flowers are pollinated by insects that are attracted to rotten flesh and have flowers that smell like dead animals, often called Carrion flowers, including *Rafflesia*, the titan arum, and the North American pawpaw (*Asimina triloba*). Flowers pollinated by night visitors, including bats and moths, are likely to concentrate on scent to attract pollinators and most such flowers are white.

Other flowers use mimicry to attract pollinators. Some species of orchids, for example, produce flowers resembling female bees in color, shape, and scent. Male bees move from one such flower to another in search of a mate.

Flower-Pollinator Relationships

Many flowers have close relationships with one or a few specific pollinating organisms. Many flowers, for example, attract only one specific species of insect, and therefore rely on that insect for successful reproduction. This close relationship is often given as an example of coevolution, as the flower and pollinator are thought to have developed together over a long period of time to match each other's needs.

This close relationship compounds the negative effects of extinction. The extinction of either member in such a relationship would mean almost certain extinction of the other member as well. Some endangered plant species are so because of shrinking pollinator populations.

Pollen Allergy

There is much confusion about the role of flowers in allergies. For example, the showy and entomophilous goldenrod (*Solidago*) is frequently blamed for respiratory allergies, of which it is innocent, since its pollen cannot be airborne. The types of pollen that most commonly cause allergic reactions are produced by the plain-looking plants (trees, grasses, and weeds) that do not have showy flowers. These plants make small, light, dry pollen grains that are custom-made for wind transport.

The type of allergens in the pollen is the main factor that determines whether the pollen is likely to cause hay fever. For example, pine tree pollen is produced in large amounts by a common tree, which would make it a good candidate for causing allergy. It is, however, a relatively rare cause of allergy because the types of allergens in pine pollen appear to make it less allergenic. Instead the allergen is usually the pollen of the contemporary bloom of anemophilous ragweed (*Ambrosia*), which can drift for many miles. Scientists have collected samples of ragweed pollen 400 miles out at sea and 2 miles high in the air. A single ragweed plant can generate a million grains of pollen per day.

Among North American plants, weeds are the most prolific producers of allergenic pollen. Ragweed is the major culprit, but other important sources are sagebrush, redroot pigweed, lamb's quarters, Russian thistle (tumbleweed), and English plantain.

It is common to hear people say they are allergic to colorful or scented flowers like roses. In fact, only florists, gardeners, and others who have prolonged, close contact with flowers are likely to be sensitive to pollen from these plants. Most people have little contact with the large, heavy, waxy pollen grains of such flowering plants because this type of pollen is not carried by wind but by insects such as butterflies and bees.

The Fruit

Fruit is the fleshy or dry ripened ovary of a flowering plant, enclosing the seed or seeds. Thus, apricots, bananas, and grapes, as well as bean pods, corn grains, tomatoes, cucumbers, and (in their shells) acorns and almonds, are all technically fruits. Popularly, however, the term is restricted to the ripened ovaries that are sweet and either succulent or pulpy.

Botanically, a fruit is a mature ovary and its associated parts. It usually contains seeds, which have developed from the enclosed ovule after fertilization, although development without fertilization, called parthenocarpy, is known, for example, in bananas. Fertilization induces various changes in a flower: the anthers and stigma wither, the petals drop off, and the sepals may be shed or undergo modifications; the ovary enlarges, and the ovules develop into seeds, each containing an embryo plant. The principal purpose of the fruit is the protection and dissemination of the seed.

Fruits are important sources of dietary fibre, vitamins (especially vitamin C), and antioxidants. Although fresh fruits are subject to spoilage, their shelf life can be extended by refrigeration or by the removal of oxygen from their storage or packaging containers. Fruits can be processed into juices, jams, and jellies and preserved by dehydration, canning, fermentation, and pickling. Waxes, such as those from bayberries (wax myrtles), and vegetable ivory from the hard fruits of a South American palm species (Phytelephas macrocarpa) are important fruit-derived products. Various drugs come from fruits, such as morphine from the fruit of the opium poppy.

Types of Fruits

The concept of "fruit" is based on such an odd mixture of practical and theoretical considerations that it accommodates cases in which one flower gives rise to several fruits (larkspur) as well as cases in which several flowers cooperate in producing one fruit (mulberry). Pea and bean plants, exemplifying the simplest situation, show in each flower a single pistil (female structure), traditionally thought of as a megasporophyll or carpel. The carpel is believed to be the evolutionary product of an originally leaf-like organ bearing ovules along its margin. This organ was somehow folded along the median line, with a meeting and coalescing of the margins of each half, the result being a miniature closed but hollow pod with one row of ovules along the suture. In many members of the rose and buttercup families, each flower contains a number of similar single-carpelled pistils, separate and distinct, which together represent what is known as an apocarpous gynoecium. In other cases, two to several carpels (still thought of as megasporophylls, although perhaps not always justifiably) are assumed to have fused to produce a single compound gynoecium (pistil), whose basal part, or ovary, may be uniloculate (with one cavity) or pluriloculate (with several compartments), depending on the method of carpel fusion.

Most fruits develop from a single pistil. A fruit resulting from the apocarpous gynoecium (several pistils) of a single flower may be referred to as an aggregate fruit. A multiple fruit represents the gynoecia of several flowers. When additional flower parts, such as the stem axis or floral tube, are retained or participate in fruit formation, as in the apple or strawberry, an accessory fruit results.

Certain plants, mostly cultivated varieties, spontaneously produce fruits in the absence of pollination and fertilization; such natural parthenocarpy leads to seedless fruits such as bananas, oranges, grapes, and cucumbers. Since 1934, seedless fruits of tomato, cucumber, peppers, holly, and others have been obtained for commercial use by administering plant growth substances, such as indoleacetic acid, indolebutyric acid, naphthalene acetic acid, and β-naphthoxyacetic acid, to the ovaries in flowers (induced parthenocarpy).

Classification systems for mature fruits take into account the number of carpels constituting the original ovary, dehiscence (opening) versus indehiscence, and dryness versus fleshiness. The properties of the ripened ovary wall, or pericarp, which may develop entirely or in part into fleshy, fibrous, or stony tissue, are important. Often three distinct pericarp layers can be identified: the outer (exocarp), the middle (mesocarp), and the inner (endocarp). All purely morphological systems (i.e., classification schemes based on structural features) are artificial. They ignore the fact that fruits can be understood only functionally and dynamically.

There are two broad categories of fruits: fleshy fruits, in which the pericarp and accessory parts develop into succulent tissues, as in eggplants, oranges, and strawberries; and dry fruits, in which the entire pericarp becomes dry at maturity. Fleshy fruits include (1) the berries, such as tomatoes, blueberries, and cherries, in which the entire pericarp and the accessory parts are succulent tissue, (2) aggregate fruits, such as blackberries and strawberries, which form from a single flower with many pistils, each of which develops into fruitlets, and (3) multiple fruits, such as pineapples and mulberries, which develop from the mature ovaries of an entire inflorescence. Dry fruits include the legumes, cereal grains, capsulate fruits, and nuts.

As strikingly exemplified by the word nut, popular terms often do not properly describe the botanical nature of certain fruits. A Brazil nut, for example, is a thick-walled seed enclosed in a likewise thick-walled capsule along with several sister seeds. A coconut is a drupe (a stony-seeded fruit) with a fibrous outer part. A walnut is a drupe in which the pericarp has differentiated into a fleshy outer husk and an inner hard "shell"; the "meat" represents the seed—two large convoluted cotyledons, a minute epicotyl and hypocotyl, and a thin papery seed coat. A peanut is an indehiscent legume fruit. An almond is a drupe "stone"; i.e., the hardened endocarp usually contains a single seed. Botanically speaking, blackberries and raspberries are not true berries but aggregates of tiny drupes. A juniper "berry" is not a fruit at all but the cone of a gymnosperm. A mulberry is a multiple fruit made up of small nutlets surrounded by fleshy sepals. And strawberry represents a much-swollen receptacle (the tip of the flower stalk bearing the flower parts) bearing on its convex surface an aggregation of tiny brown achenes (small single-seeded fruits).

Brazil nut: Hard, indehiscent fruits of the Brazil nut tree (Bertholletia excelsa). The fruit on the left has been opened to reveal the large edible seeds in their shells.

Dispersal

Fruits play an important role in the seed dispersal of many plant species. In dehiscent fruits, such as poppy capsules, the seeds are usually dispersed directly from the fruits, which may remain on the plant. In fleshy or indehiscent fruits, the seeds and fruit are commonly moved away from the parent plant together. In many plants, such as grasses and lettuce, the outer integument and ovary wall are completely fused, so seed and fruit form one entity; such seeds and fruits can logically be described together as "dispersal units," or diaspores.

Animal Dispersal

A wide variety of animals aid in the dispersal of seeds, fruits, and diaspores. Many birds and mammals, ranging in size from mice and kangaroo rats to elephants, act as dispersers when they eat fruits and diaspores. In the tropics, chiropterochory (dispersal by large bats such as flying foxes, Pteropus) is particularly important. Fruits adapted to these animals are relatively large and drab in colour with large seeds and a striking (often rank) odour. Such fruits are accessible to bats because of the pagoda-like structure of the tree canopy, fruit placement on the main trunk, or suspension from long stalks that hang free of the foliage. Examples include mangoes, guavas, breadfruit, carob, and several fig species. In South Africa a desert melon (Cucumis humifructus) participates in a symbiotic relationship with aardvarks—the animals eat the fruit for its water content and bury their own dung, which contains the seeds, near their burrows.

Cocklebur (Xanthium strumarium).

Additionally, furry terrestrial mammals are the agents most frequently involved in epizoochory, the inadvertent carrying by animals of dispersal units. Burlike fruits, or those diaspores provided with spines, hooks, claws, bristles, barbs, grapples, and prickles, are genuine hitchhikers, clinging tenaciously to their carriers. Their functional shape is achieved in various ways: in cleavers, or goose grass (Galium aparine), and in enchanter's nightshade (Circaea lutetiana), the hooks are part of the fruit itself; in common agrimony (Agrimonia eupatoria), the fruit is covered by a persistent calyx (the sepals, parts of the flower, which remain attached beyond the usual period) equipped with hooks; and in wood avens (Geum urbanum), the persistent styles have hooked tips. Other examples are bur marigolds, or beggar's-ticks (Bidens species); buffalo bur

(Solanum rostratum); burdock (Arctium); Acaena; and many Medicago species. The last-named, with dispersal units highly resistant to damage from hot water and certain chemicals (dyes), have achieved wide global distribution through the wool trade. A somewhat different principle is employed by the so-called trample burrs, said to lodge themselves between the hooves of large grazing mammals. Examples are mule grab (Proboscidea) and the African grapple plant (Harpagophytum). In water burrs, such as those of the water chestnut Trapa, the spines should probably be considered as anchoring devices.

Birds, being preening animals, rarely carry burlike diaspores on their bodies. They do, however, transport the very sticky (viscid) fruits of Pisonia, a tropical tree of the four-o'clock family, to distant Pacific islands in this way. Small diaspores, such as those of sedges and certain grasses, may also be carried in the mud sticking to waterfowl and terrestrial birds.

Chestnut-mandibled, or Swainson's,
toucan (Ramphastos swainsonii) consuming a nut.

Synzoochory, deliberate carrying of diaspores by animals, is practiced when birds carry diaspores in their beaks. The European mistle thrush (Turdus viscivorus) deposits the viscid seeds of mistletoe (Viscum album) on potential host plants when, after a meal of the berries, it whets its bill on branches or simply regurgitates the seeds. The North American (Phoradendron) and Australian (Amyema) mistletoes are dispersed by various birds, and the comparable tropical species of the plant family Loranthaceae by flower-peckers (of the bird family Dicaeidae), which have a highly specialized gizzard that allows seeds to pass through but retains insects. Plants may also profit from the forgetfulness and sloppy habits of certain nut-eating birds that cache part of their food but neglect to recover everything or that drop units on their way to a hiding place. Best known in this respect are the nutcrackers (Nucifraga), which feed largely on the "nuts" of beech, oak, walnut, chestnut, and hazelnut; the jays (Garrulus), which hide hazelnuts and acorns; the nuthatches; and the California woodpecker (Melanerpes formicivorus), which may embed literally thousands of acorns, almonds, and pecan nuts in bark fissures or holes of trees. Rodents may aid in dispersal by stealing the embedded diaspores and burying them. In Germany, an average jay may transport about 4,600 acorns per season, over distances of up to 4 km (2.5 miles).

Most ornithochores (plants with bird-dispersed seeds) have conspicuous diaspores attractive to such fruit-eating birds as thrushes, pigeons, barbets (members of the bird family Capitonidae), toucans (family Ramphastidae), and hornbills (family Bucerotidae), all of which either excrete or regurgitate the hard part undamaged. Such diaspores have a fleshy, sweet, or oil-containing edible part; a striking colour (often red or orange); no pronounced smell; protection against being eaten prematurely, in the form of acids and tannins that are present only in the green fruit; protection of the seed against digestion, afforded by bitterness, hardness, or the presence of poisonous compounds; permanent attachment; and, finally, absence of a hard outer cover. In contrast to bat-dispersed diaspores, they occupy no special position on the plant. Examples are rose hips, plums, dogwood fruits, barberry, red currant, mulberry, nutmeg fruits, figs, blackberries, and others. The natural and abundant occurrence of Euonymus, which is a largely tropical genus, in temperate Europe and Asia, can be understood only in connection with the activities of birds. Birds also contributed substantially to the repopulation with plants of the Krakatoa island group in Indonesia after the catastrophic volcanic eruption there in 1883. Birds have made Lantana (originally American) a pest in Indonesia and Australia; the same is true of black cherries (Prunus serotina) in parts of Europe, Rubus species in Brazil and New Zealand, and olives (Olea europaea) in Australia.

Bohemian waxwing (Bombycilla garrulus) eating fruit.

Many intact fruits and seeds can serve as fish bait—those of Sonneratia, for example, for the catfish Arius maculatus. Certain Amazon River fishes react positively to the audible "explosions" of the ripe fruits of Eperua rubiginosa. The largest freshwater wetlands in the world, found in Brazil's Pantanal, become inundated with seasonal floods at a time when many plants are releasing their fruits. Pacu fish (Metynnis) feed on submerged and floating fruits and disperse the seeds when they defecate. It is thought that at least one plant species (Bactris glaucescens) relies exclusively on pacu for seed dispersal.

Fossil evidence indicates that saurochory, dispersal by reptiles, is very ancient. The giant Galapagos tortoise is important for the dispersal of local cacti and tomatoes, and iguanas are known to eat and disperse a number of smaller fruits, including the iguana hackberry (Celtis iguanaea). The name alligator apple, for Annona glabra, refers to its method of dispersal, an example of saurochory.

Wind Dispersal

Winged fruits are most common in trees and shrubs, such as maple, ash, elm, birch, alder, and dipterocarps (a family of about 600 species of old world tropical trees). The one-winged propeller type, as found in maple, is called a samara. When fruits have several wings on their sides, rotation may result, as in rhubarb and dock species. Sometimes accessory parts form the wings—for example, the bracts (small green leaflike structures that grow just below flowers) in linden (Tilia).

Wind dispersal: winged fruits of the silver maple
(Acer saccharinum).

Many fruits form plumes, some derived from persisting and ultimately hairy styles, as in clematis, avens, and anemones; some from the perianth, as in the sedge family (Cyperaceae); and some from the pappus, a calyx structure, as in dandelion and Jack-go-to-bed-at-noon (Tragopogon). In woolly fruits and seeds, the pericarp or the seed coat is covered with cottonlike hairs—e.g., willow, poplar or cottonwood, cotton, and balsa. In some cases, the hairs may serve double duty in that they function in water dispersal as well as in wind dispersal.

Poppies have a mechanism in which the wind has to swing the slender fruitstalk back and forth before the seeds are thrown out through pores near the top of the capsule. The inflated indehiscent pods of Colutea arborea, a steppe plant, represent balloons capable of limited air travel before they hit the ground and become windblown tumbleweeds.

Salsify: Cluster of plumed fruits on a salsify plant (Tragopogon porrifolius).

Other Forms of Dispersal

Geocarpy is defined as either the production of fruits underground, as in the arum lilies (Stylochiton and Biarum), in which the flowers are already subterranean, or the active burying of fruits by the mother plant, as in the peanut (Arachis hypogaea). In the American hog peanut (Amphicarpa bracteata), pods of a special type are buried by the plant and are cached by squirrels later on. Kenilworth ivy (Cymbalaria), which normally grows on stone or brick walls, stashes its fruits away in crevices after strikingly extending the flower stalks. Not surprisingly, geocarpy is most often encountered in desert plants; however, it also occurs in violet species, in subterranean clover (Trifolium subterraneum), and in begonias (Begonia hypogaea) of the African rainforest.

THE ROOT SYSTEM

A root system is the network off all the roots of a plant the roots of seed plants have three major functions: anchoring the plant to the soil, absorbing water and minerals and transporting them upwards, and storing the products of photosynthesis. Some roots are modified to absorb moisture and exchange gases. Most roots are underground. Some plants, however, also have adventitious roots, which emerge above the ground from the shoot.

Types of Root Systems

Root systems are mainly of two types. Dicots have a tap root system, while monocots have a fibrous root system. A tap root system has a main root that grows down vertically, and from which many smaller lateral roots arise. Dandelions are a good example; their tap roots usually break off when trying to pull these weeds, and they can regrow another shoot from the remaining root). A tap root system penetrates deep into the soil. In contrast, a fibrous root system is located closer to the soil surface, and forms a dense network of roots that also helps prevent soil erosion (lawn grasses are a good example, as are wheat, rice, and corn). Some plants have a combination of tap roots and fibrous roots. Plants that grow in dry areas often have deep root systems, whereas plants growing in areas with abundant water are likely to have shallower root systems.

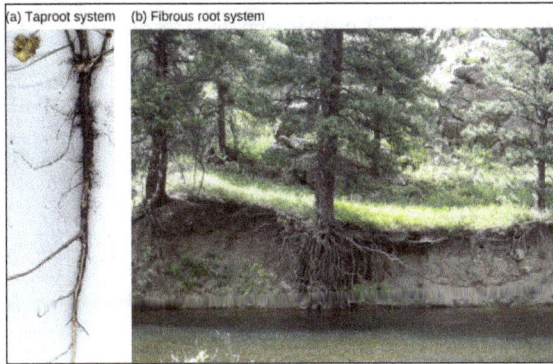

(a) Tap root systems have a main root that grows down, while (b) fibrous root systems consist of many small roots.

Root Growth and Anatomy

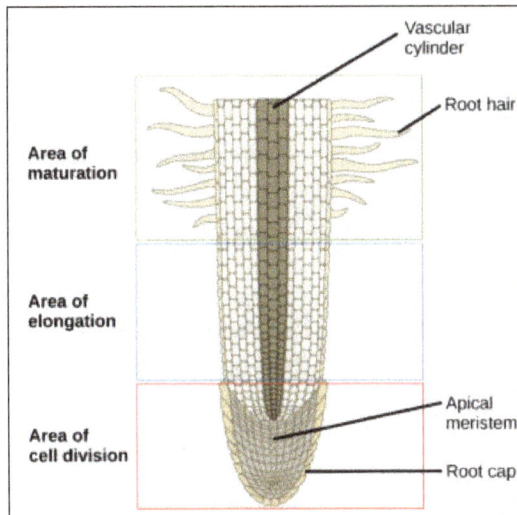

A longitudinal view of the root reveals the zones of cell division, elongation, and maturation. Cell division occurs in the apical meristem.

Root growth begins with seed germination. When the plant embryo emerges from the seed, the radicle of the embryo forms the root system. The tip of the root is protected by the root cap, a structure exclusive to roots and unlike any other plant structure. The root cap is continuously replaced because it gets damaged easily as the root pushes through soil. The root tip can be divided into three zones: a zone of cell division, a zone of elongation, and a zone of maturation and differentiation. The zone of cell division is closest to the root tip; it is made up of the actively dividing cells of the root meristem. The zone of elongation is where the newly formed cells increase in length, thereby lengthening the root. Beginning at the first root hair is the zone of cell maturation where the root cells begin to differentiate into special cell types. All three zones are in the first centimeter or so of the root tip.

The root has an outer layer of cells called the epidermis, which surrounds areas of ground tissue and vascular tissue. The epidermis provides protection and helps in absorption. Root hairs, which are extensions of root epidermal cells, increase the surface area of the root, greatly contributing to the absorption of water and minerals.

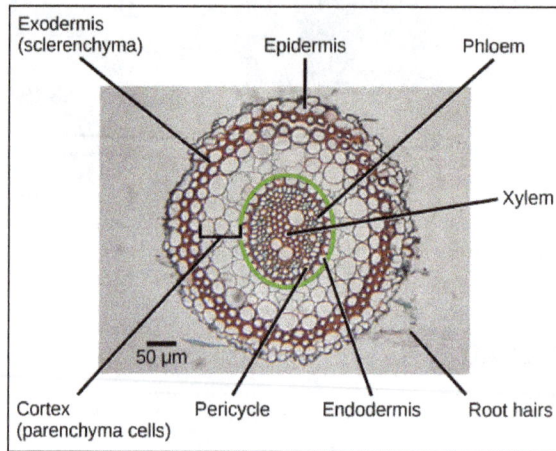

Staining reveals different cell types in this light micrograph of a wheat (Triticum) root cross section. Sclerenchyma cells of the exodermis and xylem cells stain red, and phloem cells stain blue. Other cell types stain black. The stele, or vascular tissue, is the area inside endodermis (indicated by a green ring). Root hairs are visible outside the epidermis.

Inside the root, the ground tissue forms two regions: the cortex and the pith. Compared to stems, roots have lots of cortex and little pith. Both regions include cells that store photosynthetic products. The cortex is between the epidermis and the vascular tissue, whereas the pith lies between the vascular tissue and the center of the root.

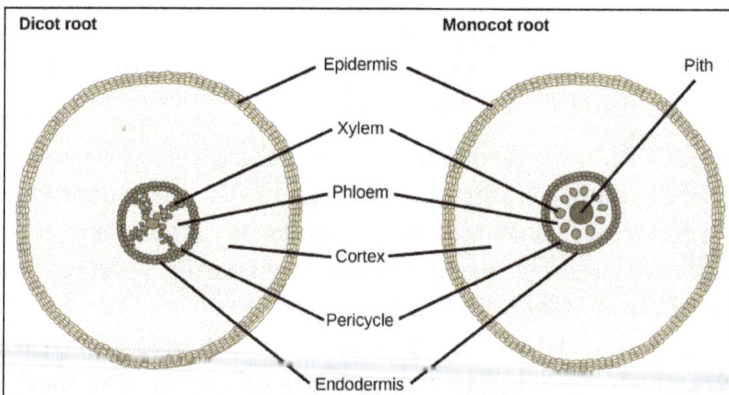

In (left) typical dicots, the vascular tissue forms an X shape in the center of the root. In (right) typical monocots, the phloem cells and the larger xylem cells form a characteristic ring around the central pith.

The vascular tissue in the root is arranged in the inner portion of the root, which is called the stele. A layer of cells known as the endodermis separates the stele from the ground tissue in the outer portion of the root. The endodermis is exclusive to roots, and serves as a checkpoint for materials entering the root's vascular system. A waxy substance called suberin is present on the walls of the endodermal cells. This waxy region, known as the Casparian strip, forces water and solutes to cross the plasma membranes of endodermal cells instead of slipping between the cells. This ensures that only materials required by the root pass through the endodermis, while toxic substances and pathogens are generally excluded. The outermost cell layer of the root's vascular tissue is the pericycle, an area that can give rise to lateral roots. In dicot roots, the xylem and phloem of the stele are arranged alternately in an X shape, whereas in monocot roots, the vascular tissue is arranged in a ring around the plth.

Root Modifications

Many vegetables are modified roots.

Root structures may be modified for specific purposes. For example, some roots are bulbous and store starch. Aerial roots and prop roots are two forms of aboveground roots that provide additional support to anchor the plant. Tap roots, such as carrots, turnips, and beets, are examples of roots that are modified for food storage.

(a)   (b)

The (a) banyan tree, also known as the strangler fig, begins life as an epiphyte in a host tree. Aerial roots extend to the ground and support the growing plant, which eventually strangles the host tree. The (b) screwpine develops aboveground roots that help support the plant in sandy soils.

Epiphytic roots enable a plant to grow on another plant. For example, the epiphytic roots of orchids develop a spongy tissue to absorb moisture. The banyan tree (Ficus sp.) begins as an epiphyte, germinating in the branches of a host tree; aerial roots develop from the branches and eventually reach the ground, providing additional support. In screwpine (Pandanus sp.), a palm-like tree that grows in sandy tropical soils, aboveground prop roots develop from the nodes to provide additional support.

The Root

Primary and secondary roots in a cotton plant.

In vascular plants, the root is the organ of a plant that typically lies below the surface of the soil. Roots can also be aerial or aerating, that is, growing up above the ground or especially above water. Furthermore, a stem normally occurring below ground is not exceptional either. Therefore, the root is best defined as the non-leaf, non-nodes bearing parts of the plant's body. However, important internal structural differences between stems and roots exist.

The first root that comes from a plant is called the radicle. A root's four major functions are:

- Absorption of water and inorganic nutrients.

- Anchoring of the plant body to the ground, and supporting it.

- Storage of food and nutrients.

- Vegetative reproduction and competition with other plants.

In response to the concentration of nutrients, roots also synthesise cytokinin, which acts as a signal as to how fast the shoots can grow. Roots often function in storage of food and nutrients. The roots of most vascular plant species enter into symbiosis with certain fungi to form mycorrhizae, and a large range of other organisms including bacteria also closely associate with roots.

Large, mature tree roots above the soil.

Anatomy

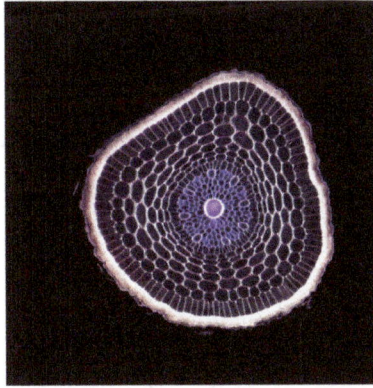

The cross-section of a barley root.

When dissected, the arrangement of the cells in a root is root hair, epidermis, epiblem, cortex, endodermis, pericycle and, lastly, the vascular tissue in the centre of a root to transport the water absorbed by the root to other places of the plant.

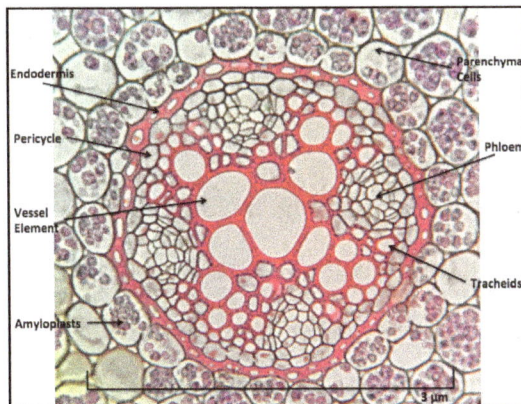

Ranunculus Root Cross Section.

Perhaps the most striking characteristic of roots (that makes it distinguishable from other plant organs such as stem-branches and leaves) is that, roots have an endogenous origin, i.e. it originates and develops from an inner layer of the mother axis (Such as Pericycle). Whereas Stem-branching and leaves (those develop as buds) are exogenous, i.e. start to develop from the cortex, an outer layer.

Architecture

Tree roots at Cliffs of the Neuse State Park.

In its simplest form, the term root architecture refers to the spatial configuration of a plant's root system. This system can be extremely complex and is dependent upon multiple factors such as the species of the plant itself, the composition of the soil and the availability of nutrients.

The configuration of root systems serves to structurally support the plant, compete with other plants and for uptake of nutrients from the soil. Roots grow to specific conditions, which, if changed, can impede a plant's growth. For example, a root system that has developed in dry soil may not be as efficient in flooded soil, yet plants are able to adapt to other changes in the environment, such as seasonal changes.

Root architecture plays the important role of providing a secure supply of nutrients and water as well as anchorage and support. The main terms used to classify the architecture of a root system are:

- Branch magnitude: the number of links (exterior or interior).

- Topology: the pattern of branching, including:

  ○ Herringbone: alternate lateral branching off a parent root.

  ○ Dichotomous: opposite, forked branches.

  ○ Radial: whorl(s) of branches around a root.

- Link length: the distance between branches.

- Root angle: the radial angle of a lateral root's base around the parent root's circumference, the angle of a lateral root from its parent root, and the angle an entire system spreads.

- Link radius: the diameter of a root.

All components of the root architecture are regulated through a complex interaction between genetic responses and responses due to environmental stimuli. These developmental stimuli are categorised as intrinsic, the genetic and nutritional influences, or extrinsic, the environmental influences and are interpreted by signal transduction pathways. The extrinsic factors that affect root architecture include gravity, light exposure, water and oxygen, as well as the availability or lack of nitrogen, phosphorus, sulphur, aluminium and sodium chloride. The main hormones (intrinsic stimuli) and respective pathways responsible for root architecture development include:

- Auxin: Auxin promotes root initiation, root emergence and primary root elongation.

- Cytokinins: Cytokinins regulate root apical meristem size and promote lateral root elongation.

- Gibberellins: Together with ethylene they promote crown primordia growth and elongation. Together with auxin they promote root elongation. Gibberellins also inhibit lateral root primordia initiation.

- Ethylene: Ethylene promotes crown root formation.

Growth

Roots of trees.

Early root growth is one of the functions of the apical meristem located near the tip of the root. The meristem cells more or less continuously divide, producing more meristem, root cap cells (these are sacrificed to protect the meristem), and undifferentiated root cells. The latter become the primary tissues of the root, first undergoing elongation, a process that pushes the root tip forward in the growing medium. Gradually these cells differentiate and mature into specialized cells of the root tissues.

Growth from apical meristems is known as primary growth, which encompasses all elongation. Secondary growth encompasses all growth in diameter, a major component of woody plant tissues and many nonwoody plants. For example, storage roots of sweet potato have secondary growth but are not woody. Secondary growth occurs at the lateral meristems, namely the vascular cambium and cork cambium. The former forms secondary xylem and secondary phloem, while the latter forms the periderm.

In plants with secondary growth, the vascular cambium, originating between the xylem and the phloem, forms a cylinder of tissue along the stem and root.The vascular cambium forms new cells on both the inside and outside of the cambium cylinder, with those on the inside forming secondary xylem cells, and those on the outside forming secondary phloem cells. As secondary xylem accumulates, the "girth" (lateral dimensions) of the stem and root increases. As a result, tissues beyond the secondary phloem including the epidermis and cortex, in many cases tend to be pushed outward and are eventually "sloughed off" (shed).

At this point, the cork cambium begins to form the periderm, consisting of protective cork cells containing suberin. In roots, the cork cambium originates in the pericycle, a component of the vascular cylinder.

The vascular cambium produces new layers of secondary xylem annually. The xylem vessels are dead at maturity but are responsible for most water transport through the vascular tissue in stems and roots.

Tree roots usually grow to three times the diameter of the branch spread, only half of which lie underneath the trunk and canopy. The roots from one side of a tree usually supply nutrients to the foliage on the same side. Some families however, such as Sapindaceae (the maple family), show no correlation between root location and where the root supplies nutrients on the plant.

Regulation

There is a correlation of roots using the process of plant perception to sense their physical environment to grow, including the sensing of light, and physical barriers. Plants also sense gravity and respond through auxin pathways, resulting in gravitropism. Over time, roots can crack foundations, snap water lines, and lift sidewalks. Research has shown that roots have ability to recognize 'self' and 'non-self' roots in same soil environment.

The correct environment of air, mineral nutrients and water directs plant roots to grow in any direction to meet the plant's needs. Roots will shy or shrink away from dry or other poor soil conditions.

Gravitropism directs roots to grow downward at germination, the growth mechanism of plants that also causes the shoot to grow upward.

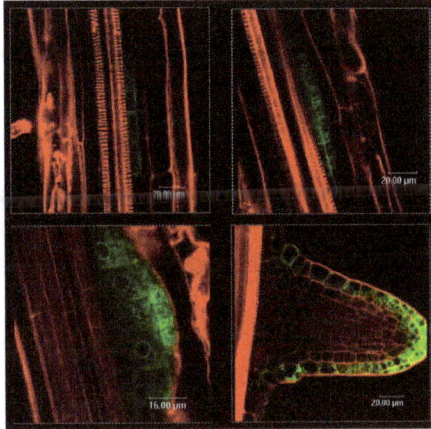

Fluorescent imaging of an emerging lateral root.

Shade Avoidance Root Response

In order to avoid shade, plants utilize a shade avoidance response. When a plant is under dense vegetation, the presence of other vegetation nearby will cause the plant to avoid lateral growth and experience an increase in upward shoot, as well as downward root growth. In order to escape shade, plants adjust their root architecture, most notably by decreasing the length and amount of lateral roots emerging from the primary root. Experimentation of mutant variants of Arabidospis thaliana found that plants sense the Red to Far Red light ratio that enters the plant through photoreceptors known as phytochromes. Nearby plant leaves will absorb red light and reflect far- red light which will cause the ratio red to far red light to lower.

The phytochrome PhyA that senses this Red to Far Red light ratio is localized in both the root system as well as the shoot system of plants, but through knockout mutant experimentation, it was found that root localized PhyA does not sense the light ratio, whether directly or axially, that leads to changes in the lateral root architecture. Research instead found that shoot localized PhyA is the phytochrome responsible for causing these architectural changes of the lateral root. Research has also found that phytochrome completes these architectural changes through the manipulation of auxin distribution in the root of the plant. When a low enough Red to Far Red ratio is sensed by PhyA, the phyA in the shoot will be mostly in its active form. In this form, PhyA stabilize the transcription factor HY5 causing it to no longer be degraded as it is when phyA is in its inactive form. This stabilized transcription factor is then able to be transported to the roots of the plant through the phloem, where it proceeds to induce its own transcription as a way to amplify its signal. In the roots of the plant HY5 functions to inhibit an auxin response factor known as ARF19, a response factor responsible for the translation of PIN3 and LAX3, two well known auxin transporting proteins. Thus, through manipulation of ARF19, the level and activity of auxin transporters PIN3 and LAX3 is inhibited. Once inhibited, auxin levels will be low in areas where lateral root emergence normally occurs, resulting in a failure for the plant to have the emergence of the lateral root primordium through the root pericycle. With this complex manipulation of Auxin transport in the roots, lateral root emergence will be inhibited in the roots and the root will instead elongate downwards, promoting vertical plant growth in an attempt to avoid shade.

Research of Arabidopsis has led to the discovery of how this auxin mediated root response works. In an attempt to discover the role that phytochrome plays in lateral root development, Salisbury et al. worked with Arabidopsis thaliana grown on agar plates. Salisbury et al. used wild type plants along with varying protein knockout and gene knockout Arabidopsis mutants to observe the results these mutations had on the root architecture, protein presence, and gene expression. To do this, Salisbury et al. used GFP fluorescence along with other forms of both macro and microscopic imagery to observe any changes various mutations caused. From these research, Salisbury et al. were able to theorize that shoot located phytochromes alter auxin levels in roots, controlling lateral root development and overall root architecture. In the experiments of van Gelderen et al., they wanted to see if and how it is that the shoot of Arabidopsis thaliana alters and affects root development and root architecture. To do this, they took Arabidopsis plants, grew them in agar gel, and exposed the roots and shoots to separate sources of light. From here, they altered the different wavelengths of light the shoot and root of the plants were receiving and recorded the lateral root density, amount of lateral roots, and the general architecture of the lateral roots. To identify the function of specific photoreceptors, proteins, genes, and hormones, they utilized various Arabidopsis knockout mutants and observed the resulting changes in lateral roots architecture. Through their observations and various experiments, van Gelderen et al. were able to develop a mechanism for how root detection of Red to Far-red light ratios alter lateral root development.

Type

A true root system consists of a primary root and secondary roots (or lateral roots).

- The diffuse root system: the primary root is not dominant; the whole root system is fibrous and branches in all directions. Most common in monocots. The main function of the fibrous root is to anchor the plant.

Specialized

The roots, or parts of roots, of many plant species have become specialized to serve adaptive purposes besides the two primary functions described in the introduction.

- Adventitious roots arise out-of-sequence from the more usual root formation of branches of a primary root, and instead originate from the stem, branches, leaves, or old woody roots. They commonly occur in monocots and pteridophytes, but also in many dicots, such as clover (*Trifolium*), ivy (*Hedera*), strawberry (*Fragaria*) and willow (*Salix*). Most aerial roots and stilt roots are adventitious. In some conifers adventitious roots can form the largest part of the root system.

- Aerating roots (or knee root or knee or pneumatophores): roots rising above the ground, especially above water such as in some mangrove genera (*Avicennia*, *Sonneratia*). In some plants like *Avicennia* the erect roots have a large number of breathing pores for exchange of gases.

- Aerial roots: roots entirely above the ground, such as in ivy (*Hedera*) or in epiphytic orchids. Many aerial roots are used to receive water and nutrient intake directly from the air - from fogs, dew or humidity in the air. Some rely on leaf systems to gather rain or humidity and even store it in scales or pockets. Other aerial roots, such as mangrove aerial roots, are used for aeration and not for water absorption. Other aerial roots are used mainly for structure, functioning as prop roots, as in maize or anchor roots or as the trunk in strangler fig. In some Epiphytes - plants living above the surface on other plants, aerial roots serve for reaching to water sources or reaching the surface, and then functioning as regular surface roots.

- Canopy Roots/Arboreal Roots: forms when tree branches support mats of epiphytes and detritus, which hold water and nutrients in the canopy. Tree branches send out canopy roots into these mats, likely to utilize the available nutrients and moisture.

- Contractile roots: these pull bulbs or corms of monocots, such as hyacinth and lily, and some taproots, such as dandelion, deeper in the soil through expanding radially and contracting longitudinally. They have a wrinkled surface.

- Coarse roots: roots that have undergone secondary thickening and have a woody structure. These roots have some ability to absorb water and nutrients, but their main function is transport and to provide a structure to connect the smaller diameter, fine roots to the rest of the plant.

- Dimorphic root systems: roots with two distinctive forms for two separate functions.

- Fine roots: typically primary roots <2 mm diameter that have the function of water and nutrient uptake. They are often heavily branched and support mycorrhizas. These roots may be short lived, but are replaced by the plant in an ongoing process of root 'turnover'.

- Haustorial roots: roots of parasitic plants that can absorb water and nutrients from another plant, such as in mistletoe (*Viscum album*) and dodder.

- Propagative roots: roots that form adventitious buds that develop into aboveground shoots, termed suckers, which form new plants, as in Canada thistle, cherry and many others.

- Proteoid roots or cluster roots: dense clusters of rootlets of limited growth that develop under low phosphate or low iron conditions in Proteaceae and some plants from the following families Betulaceae, Casuarinaceae, Elaeagnaceae, Moraceae, Fabaceae and Myricaceae.

- Stilt roots: these are adventitious support roots, common among mangroves. They grow down from lateral branches, branching in the soil.

- Storage roots: these roots are modified for storage of food or water, such as carrots and beets. They include some taproots and tuberous roots.

- Structural roots: large roots that have undergone considerable secondary thickening and provide mechanical support to woody plants and trees.

- Surface roots: these proliferate close below the soil surface, exploiting water and easily available nutrients. Where conditions are close to optimum in the surface layers of soil, the growth of surface roots is encouraged and they commonly become the dominant roots.

- Tuberous roots: fleshy and enlarged lateral roots for food or water storage, e.g. sweet potato. A type of storage root distinct from taproot.

Stilt roots of Maize plant.

Aerial root.

Aerating roots of a mangrove.

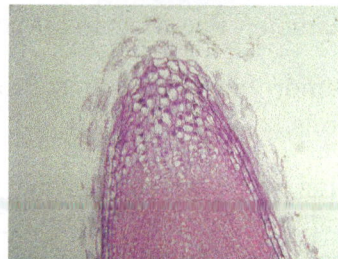
The growing tip of a fine root.

Visible roots.

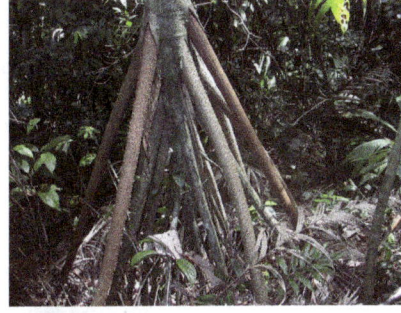
The stilt roots of *Socratea exorrhiza*.

Depths

The distribution of vascular plant roots within soil depends on plant form, the spatial and temporal availability of water and nutrients, and the physical properties of the soil. The deepest roots are generally found in deserts and temperate coniferous forests; the shallowest in tundra, boreal forest and temperate grasslands. The deepest observed living root, at least 60 metres below the ground surface, was observed during the excavation of an open-pit mine in Arizona, USA. Some roots can grow as deep as the tree is high. The majority of roots on most plants are however found relatively close to the surface where nutrient availability and aeration are more favourable for growth. Rooting depth may be physically restricted by rock or compacted soil close below the surface, or by anaerobic soil conditions.

Cross section of a mango tree.

Depth Records

Species	Location	Maximum rooting depth (m)
Boscia albitrunca	Kalahari desert	68
Juniperus monosperma	Colorado Plateau	61
Eucalyptus sp.	Australian forest	61

Acacia erioloba	Kalahari desert	60
Prosopis juliflora	Arizona desert	53.3

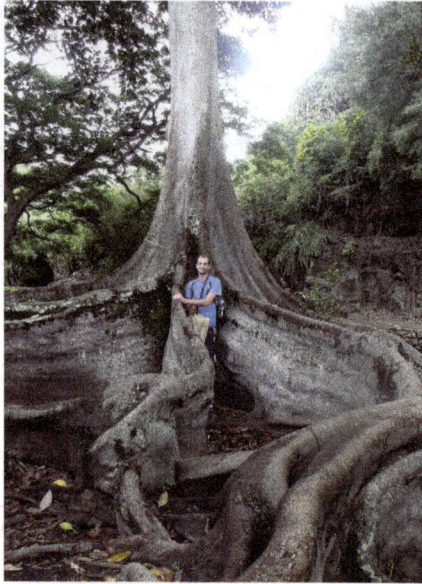

Ficus Tree with buttress roots.

Environmental Interactions

Certain plants, namely Fabaceae, form root nodules in order to associate and form a symbiotic relationship with nitrogen-fixing bacteria called rhizobia. Due to the high energy required to fix nitrogen from the atmosphere, the bacteria take carbon compounds from the plant to fuel the process. In return, the plant takes nitrogen compounds produced from ammonia by the bacteria.

Plants can also interact with one another in their environment through their root systems. Studies have demonstrated that plant-plant interaction occurs among root systems via the soil as a medium. For instance, Novoplanksy and his students at Ben-Gurion University in Israel tested whether plants growing in ambient conditions would change their behavior if a nearby plant was exposed to drought conditions. His team wondered if plants can communicate to their neighbors of nearby stressful environmental conditions.

To investigate this, Novoplansky and his team set up a "split root" experimental design, in which a plant's roots were split between two pots (say, Pot A and Pot B). A second plant's roots were then placed between two pots, such that half its roots were in the same pot as Plant 1 (Pot B), and half its roots in a new pot (Pot C). A third plant was chained to the first two plants, in which plant 3's roots were split between sharing Pot C with the second plant's roots and a new Pot D, and so on. In the study, six pea plants (*Pisum sativum*) were chained together in seven pots.

Drought was stimulated by adding mannitol, a sugar commonly used to initiate drought responses in plant studies, to Pot A such that only half the roots of the plant 1 were subjected to drought conditions. The other six pots remained well watered and weren't exposed to mannitol. After adding mannitol to the soil of Pot A, that plant 1's stomata closed within fifteen minutes even though half its roots remained in the well-watered soil of Pot B. Novoplansky's team collected stomatal aperture data using epidermal impressions. This result wasn't surprising, since stomatal closing as a drought response has been widely studied; in this response, Abscisic Acid (ABA) is produced in the roots and sent up to leaves to close the stomates. However, the stomata of plant 2, with half its roots sharing Pot B with half the roots of the first plant, also closed within 15 minutes of mannitol addition to Pot A. This observation suggested that a signal from the drought-induced roots in Pot A travelled to the non-stressed roots of the same plant, which caused them to release a drought signal into Pot B's soil.

Stomatal closing at the leaves of the other plant neighbors in the remaining pots was also measured, albeit at a longer time after the mannitol treatment in Pot A. These results implied that the drought signal was relayed to plants five pots away from the stress origin. Neighboring plants placed in separate pots directly next to Pot A showed no changes in stomatal aperture, which confirmed that the drought signal spread through the roots and soil, not through the air as a volatile chemical signal. Novoplansky and his team concluded that information about environmental conditions must be communicated through root systems, implying that plant communication occurs among roots of the same plant and nearby plants as well. Further research is underway as to how the plants sense drought and respond by emitting a chemical drought signal into the soil to alert nearby roots of stressful conditions. As of now, the chemical signal is unknown, as is the physiological mechanism by which this response occurs.

Economic Importance

Roots can also protect the environment by holding the soil
to reduce soil erosion.

The term root crops refers to any edible underground plant structure, but many root crops are actually stems, such as potato tubers. Edible roots include cassava, sweet potato, beet, carrot, rutabaga, turnip, parsnip, radish, yam and horseradish. Spices obtained from roots include sassafras, angelica, sarsaparilla and licorice.

Sugar beet is an important source of sugar. Yam roots are a source of estrogen compounds used in birth control pills. The fish poison and insecticide rotenone is obtained from roots of *Lonchocarpus* spp. Important medicines from roots are ginseng, aconite, ipecac, gentian and reserpine. Several legumes that have nitrogen-fixing root nodules are used as green manure crops, which provide nitrogen fertilizer for other crops when plowed under. Specialized bald cypress roots, termed knees, are sold as souvenirs, lamp bases and carved into folk art. Native Americans used the flexible roots of white spruce for basketry.

Tree roots can heave and destroy concrete sidewalks and crush or clog buried pipes. The aerial roots of strangler fig have damaged ancient Mayan temples in Central America and the temple of Angkor Wat in Cambodia.

Trees stabilize soil on a slope prone to landslides. The root hairs work as an anchor on the soil.

Vegetative propagation of plants via cuttings depends on adventitious root formation. Hundreds of millions of plants are propagated via cuttings annually including chrysanthemum, poinsettia, carnation, ornamental shrubs and many houseplants.

Roots can also protect the environment by holding the soil to reduce soil erosion. This is especially important in areas such as sand dunes.

Plant Stem

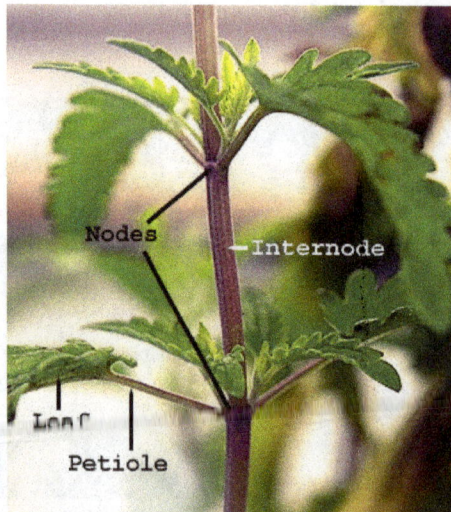

Stem showing internode and nodes plus leaf petioles.

This above-ground stem of *Polygonatum* has lost its leaves, but is producing adventitious roots from the nodes.

A stem is one of two main structural axes of a vascular plant, the other being the root. The stem is normally divided into nodes and internodes:

- The nodes hold one or more leaves, as well as buds which can grow into branches (with leaves, conifer cones, or inflorescences (flowers)). Adventitious roots may also be produced from the nodes.

- The internodes distance one node from another.

The term "shoots" is often confused with "stems"; "shoots" generally refers to new fresh plant growth including both stems and other structures like leaves or flowers. In most plants stems are located above the soil surface but some plants have underground stems.

Stems have four main functions which are:

- Support for and the elevation of leaves, flowers and fruits. The stems keep the leaves in the light and provide a place for the plant to keep its flowers and fruits.

- Transport of fluids between the roots and the shoots in the xylem and phloem.

- Storage of nutrients.

- Production of new living tissue. The normal lifespan of plant cells is one to three years. Stems have cells called meristems that annually generate new living tissue. Specialized terms.

Stems are often specialized for storage, asexual reproduction, protection or photosynthesis, including the following:

- Acaulescent – used to describe stems in plants that appear to be stemless.

Actually these stems are just extremely short, the leaves appearing to rise directly out of the ground, e.g. some *Viola* species.

- Arborescent – tree like with woody stems normally with a single trunk.

- Axillary bud – a bud which grows at the point of attachment of an older leaf with the stem. It potentially gives rise to a shoot.

- Branched – aerial stems are described as being branched or unbranched.

- Bud – an embryonic shoot with immature stem tip.

- Bulb – a short vertical underground stem with fleshy storage leaves attached, e.g. onion, daffodil, tulip. Bulbs often function in reproduction by splitting to form new bulbs or producing small new bulbs termed bulblets. Bulbs are a combination of stem and leaves so may better be considered as leaves because the leaves make up the greater part.

- Caespitose – when stems grow in a tangled mass or clump or in low growing mats.

- Cladode (including phylloclade) – a flattened stem that appears more-or-less leaf like and is specialized for photosynthesis, e.g. cactus pads.

- Climbing – stems that cling or wrap around other plants or structures.

- Corm – a short enlarged underground, storage stem, e.g. taro, crocus, gladiolus.

Decumbent stem in *Cucurbita maxima*:

- Decumbent – stems that lie flat on the ground and turn upwards at the ends.

- Fruticose – stems that grow shrublike with woody like habit.

- Herbaceous – non woody, they die at the end of the growing season.

- Internode – an interval between two successive nodes. It possesses the ability to elongate, either from its base or from its extremity depending on the species.

- Node – a point of attachment of a leaf or a twig on the stem in seed plants. A node is a very small growth zone.

- Pedicel – stems that serve as the stalk of an individual flower in an inflorescence or infrutescence.

- Peduncle – a stem that supports an inflorescence.

- Prickle – a sharpened extension of the stem's outer layers, e.g. roses.

- Pseudostem – a false stem made of the rolled bases of leaves, which may be 2 or 3 m tall as in banana.

- Rhizome – a horizontal underground stem that functions mainly in reproduction but also in storage, e.g. most ferns, iris.

- Runner (plant part) – a type of stolon, horizontally growing on top of the ground and rooting at the nodes, aids in reproduction. e.g. garden strawberry, *Chlorophytum comosum*.

- Scape – a stem that holds flowers that comes out of the ground and has no normal leaves. Hosta, Lily, Iris, Garlic.

- Stolon – a horizontal stem that produces rooted plantlets at its nodes and ends, forming near the surface of the ground.

- Thorn – a modified stem with a sharpened point.

- Tuber – a swollen, underground storage stem adapted for storage and reproduction, e.g. potato.

- Woody – hard textured stems with secondary xylem.

Stem Structure

Stem usually consist of three tissues, dermal tissue, ground tissue and vascular tissue. The dermal tissue covers the outer surface of the stem and usually functions to waterproof, protect and control gas exchange. The ground tissue usually consists mainly of parenchyma cells and fills in around the vascular tissue. It sometimes functions in photosynthesis. Vascular tissue provides long distance transport and structural support. Most or all ground tissue may be lost in woody stems. The dermal tissue of aquaticplants stems may lack the waterproofing found in aerial stems. The arrangement of the vascular tissues varies widely among plant species.

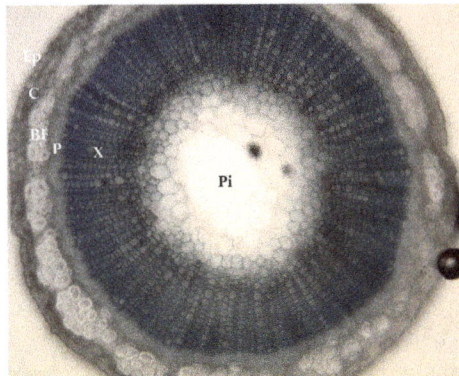

Flax stem cross-section, showing locations of underlying tissues. Ep = epidermis; C = cortex; BF = bast fibres; P = phloem; X = xylem; Pi = pith.

Dicot Stems

Dicot stems with primary growth have pith in the center, with vascular bundles forming a distinct ring visible when the stem is viewed in cross section. The outside of the stem is covered with an epidermis, which is covered by a waterproof cuticle. The epidermis also may contain stomata for gas exchange and multicellular stem hairs called trichomes. A cortex consisting of hypodermis (collenchyma cells) and endodermis (starch containing cells) is present above the pericycle and vascular bundles.

Woody dicots and many nonwoody dicots have secondary growth originating from their lateral or secondary meristems: the vascular cambium and the cork cambium or phellogen. The vascular cambium forms between the xylem and phloem in the vascular bundles and connects to form a continuous cylinder. The vascular cambium cells divide to produce secondary xylem to the inside and secondary phloem to the outside. As the stem increases in diameter due to production of secondary xylem and secondary phloem, the cortex and epidermis are eventually destroyed. Before the cortex is destroyed, a cork cambium develops there. The cork cambium divides to produce waterproof cork cells externally and sometimes phelloderm cells internally. Those three tissues form the periderm, which replaces the epidermis in function. Areas of loosely packed cells in the periderm that function in gas exchange are called lenticels.

Secondary xylem is commercially important as wood. The seasonal variation in growth from the vascular cambium is what creates yearly tree rings in temperate climates. Tree rings are the basis of dendrochronology, which dates wooden objects and associated artifacts. Dendroclimatology is the use of tree rings as a record of past climates. The aerial stem of an adult tree is called a trunk. The dead, usually darker inner wood of a large diameter trunk is termed the heartwood and is the result of tylosis. The outer, living wood is termed the sapwood.

Monocot Stems

Stems of two Roystonea regia palms showing characteristic bulge,
leaf scars and fibrous roots, Kolkata, India.

Vascular bundles are present throughout the monocot stem, although concentrated towards the outside. This differs from the dicot stem that has a ring of vascular bundles and often none in the center. The shoot apex in monocot stems is more elongated. Leaf sheathes grow up around it, protecting it. This is true to some extent of almost all monocots. Monocots rarely produce secondary growth and are therefore seldom woody, with Palms and Bamboo being notable exceptions. However, many monocot stems increase in diameter via anomalous secondary growth.

Gymnosperm Stems

All gymnosperms are woody plants. Their stems are similar in structure to woody dicots except that most gymnosperms produce only tracheids in their xylem, not the vessels found in dicots. Gymnosperm wood also often contains resin ducts. Woody dicots are called hardwoods, e.g. oak, maple and walnut. In contrast, softwoods are gymnosperms, such as pine, spruce and fir.

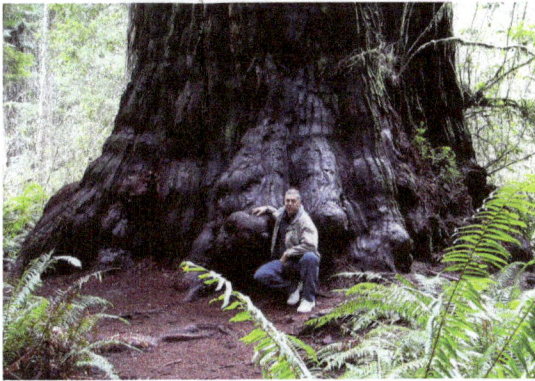

The trunk of this redwood tree is its stem.

Tasmanian tree fern.

Fern Stems

Most ferns have rhizomes with no vertical stem. The exception is tree ferns, with vertical stems up to about 20 metres. The stem anatomy of ferns is more complicated than that of dicots because fern stems often have one or more leaf gaps in cross section. A leaf gap is where the vascular tissue branches off to a frond. In cross section, the vascular tissue does not form a complete cylinder where a leaf gap occurs. Fern stems may have solenosteles or dictyosteles or variations of them. Many fern stems have phloem tissue on both sides of the xylem in cross-section.

Relation to Xenobiotics

Foreign chemicals such as air pollutants, herbicides and pesticides can damage stem structures.

Economic Importance

White and green asparagus – crispy stems are the edible parts of this vegetable.

There are thousands of species whose stems have economic uses. Stems provide a few major staple crops such as potato and taro. Sugarcane stems are a major source of sugar. Maple sugar is obtained from trunks of maple trees. Vegetables from stems are asparagus, bamboo shoots, cactus pads or nopalitos, kohlrabi, and water chestnut. The spice, cinnamon is bark from a tree trunk. Gum arabic is an important food additive obtained from the trunks of *Acacia senegal* trees. Chicle, the main ingredient in chewing gum, is obtained from trunks of the chicle tree.

Medicines obtained from stems include quinine from the bark of cinchona trees, camphor distilled from wood of a tree in the same genus that provides cinnamon, and the muscle relaxant curare from the bark of tropical vines.

Wood is used in thousands of ways, e.g. buildings, furniture, boats, airplanes, wagons, car parts, musical instruments, sports equipment, railroad ties, utility poles, fence posts, pilings, toothpicks, matches, plywood, coffins, shingles, barrel staves, toys, tool handles, picture frames, veneer, charcoal and firewood. Wood pulp is widely used to make paper, paperboard, cellulose sponges, cellophane and some important plastics and textiles, such as cellulose acetate and rayon. Bamboo stems also have hundreds of uses, including paper, buildings, furniture, boats, musical instruments, fishing poles, water pipes, plant stakes, and scaffolding. Trunks of palm trees and tree ferns are often used for building. Stems of Reed are an important building material for use in thatching in some areas.

Tannins used for tanning leather are obtained from the wood of certain trees, such as quebracho. Cork is obtained from the bark of the cork oak. Rubber is obtained from the trunks of *Hevea brasiliensis*. Rattan, used for furniture and baskets, is made from the stems of tropical vining palms. Bast fibers for textiles and rope are obtained from stems include flax, hemp, jute and ramie. The earliest paper was obtained from the stems of papyrus by the ancient Egyptians.

Amber is fossilized sap from tree trunks; it is used for jewelry and may contain ancient animals. Resins from conifer wood are used to produce turpentine and rosin. Tree bark is often used as a mulch and in growing media for container plants. It also can become the natural habitat of lichens.

Some ornamental plants are grown mainly for their attractive stems, e.g.:

- White bark of paper birch.

- Twisted branches of corkscrew willow and Harry Lauder's walking stick (Corylus avellana 'Contorta').

- Red, peeling bark of paperbark maple.

References

- Plant-anatomy: quinessence.com, Retrieved 3 May, 2019

- Leins, P. & Erbar, C. (2010). Flower and Fruit. Stuttgart: Schweizerbart Science Publishers. ISBN 978-3-510-65261-7

- Plant-shoot-system-structure-function-quiz, lesson, academy: study.com, Retrieved 15 July, 2019

- De Craene, Louis P. Ronse (2010). Floral Diagrams. Cambridge University Press. P. 459. ISBN 9781139484558

- Biobookplantanat, biobk, farabee, faculty: estrellamountain.edu, Retrieved 5 February, 2019

- Sharma, O.P. (2009). Plant Taxonomy (2nd ed.). Tata mcgraw-Hill Education. Pp. 165–166. ISBN 978-1259081378. Archived from the original on 2016-05-29

- Fruit-plant-reproductive-body, science: britannica.com, Retrieved 16 January, 2019

- Malamy JE (2005). "Intrinsic and environmental response pathways that regulate root system architecture". Plant, Cell & Environment. 28: 67–77. Doi:10.1111/j.1365-3040.2005.01306.x

- Roots, chapter, wm-biology2: lumenlearning.com Retrieved 29 March, 2019

- Malamy JE, Ryan KS (November 2001). "Environmental regulation of lateral root initiation in Arabidopsis". Plant Physiology. 127 (3): 899–909. Doi:10.1104/pp.010406. PMC 129261. PMID 11706172

- Stem-plant, science: britannica.com, Retrieved 30 August, 2019

- Raven, Peter H., Ray Franklin Evert, and Helena Curtis. 1981. Biology of plants. New York, N.Y Worth Publishers. ISBN 0-87901-132-7

PERMISSIONS

All chapters in this book are published with permission under the Creative Commons Attribution Share Alike License or equivalent. Every chapter published in this book has been scrutinized by our experts. Their significance has been extensively debated. The topics covered herein carry significant information for a comprehensive understanding. They may even be implemented as practical applications or may be referred to as a beginning point for further studies.

We would like to thank the editorial team for lending their expertise to make the book truly unique. They have played a crucial role in the development of this book. Without their invaluable contributions this book wouldn't have been possible. They have made vital efforts to compile up to date information on the varied aspects of this subject to make this book a valuable addition to the collection of many professionals and students.

This book was conceptualized with the vision of imparting up-to-date and integrated information in this field. To ensure the same, a matchless editorial board was set up. Every individual on the board went through rigorous rounds of assessment to prove their worth. After which they invested a large part of their time researching and compiling the most relevant data for our readers.

The editorial board has been involved in producing this book since its inception. They have spent rigorous hours researching and exploring the diverse topics which have resulted in the successful publishing of this book. They have passed on their knowledge of decades through this book. To expedite this challenging task, the publisher supported the team at every step. A small team of assistant editors was also appointed to further simplify the editing procedure and attain best results for the readers.

Apart from the editorial board, the designing team has also invested a significant amount of their time in understanding the subject and creating the most relevant covers. They scrutinized every image to scout for the most suitable representation of the subject and create an appropriate cover for the book.

The publishing team has been an ardent support to the editorial, designing and production team. Their endless efforts to recruit the best for this project, has resulted in the accomplishment of this book. They are a veteran in the field of academics and their pool of knowledge is as vast as their experience in printing. Their expertise and guidance has proved useful at every step. Their uncompromising quality standards have made this book an exceptional effort. Their encouragement from time to time has been an inspiration for everyone.

The publisher and the editorial board hope that this book will prove to be a valuable piece of knowledge for students, practitioners and scholars across the globe.

INDEX

A

Abscission, 91, 95, 115

Acritarchs, 7, 9

Agrobacterium Tumefaciens, 27

Anaerobic Soil, 173

Anemophilous Flowers, 149

Angiosperms, 7, 51, 53-54, 57, 64, 76, 78, 102, 108-109, 113, 126, 129, 133-134, 136, 140, 142

Anthocyanins, 89-90, 115

Antirrhinum Majus, 79, 148

Apical Meristems, 74-75, 77, 79-80, 168

Arabidopsis Thaliana, 26, 42, 75,78, 91, 93, 148, 170

Axillary Buds, 96

B

Biennial Plants, 82, 141, 147

Biomass, 20, 22

Biotic Interactions, 19

Bryology, 6, 29-32

Bryophytes, 1, 29-31, 102

Bryum Cryophyllum, 29

C

Calyx, 83, 143, 145, 156, 159

Campylodromous, 125, 127

Canopy Roots, 171

Central Vacuole, 35, 47

Chitinozoans, 7, 9

Chlorenchyma, 47, 62, 67, 111-112

Chloroplast, 24, 35, 39, 44, 71, 111

Citrus Sinensis, 83

Collenchyma Cells, 66, 68, 71, 180

Cross-fertilization, 23

D

Dicot Stems, 179

Dicotyledons, 107, 126, 140

Dinoflagellate Cysts, 7, 9

E

Epidermal Cells, 48, 65, 69, 111, 162

Ethnobotany, 6, 10-15, 32

F

Fibrous Root System, 160

G

Gene Gun, 28

Genetically Modified Crops, 27

Ginkgo Biloba, 133

H

Hickey System, 126, 129-130

Hornworts, 29, 53

Hybrid Vigor, 25

L

Lavandula Angustifolia, 83

Leaf Blades, 76, 94, 107, 114

Liverworts, 29-31, 98, 102

M

Meristematic Tissue, 46

Mesophyll Tissue, 111

Mitochondria, 24-25, 36, 44, 62

Mutualism, 17, 21, 94

Mycorrhizae, 21, 164

Myrtaceae, 64, 137

N

Nicotiana Benthamiana, 27

P

Paleobotany, 6-7, 32

Palisade Mesophyll, 38, 97, 113

Parenchyma Tissue, 71, 113

Particulate Organic Matter, 7-8

Perennial Plants, 82, 141

Petiole, 84, 96, 98, 102-103, 105-109, 114, 131-132

Phloem, 36-38, 47-50, 58, 60-65, 70-72, 79, 99, 110, 113, 140-141, 162-163, 168-169, 177, 179-181

Phormium Tenax, 69-70

Photoperiodism, 85, 87, 92

Phytelephas Macrocarpa, 154

Phytohormones, 91

Phytoliths, 6, 101, 116

Pinnate, 84, 105-107, 125-126, 129, 131-134, 136, 139

Plant Ecology, 6, 16-19, 32, 85, 93

Plant Growth Regulators, 77, 91

Poinsettia, 92-93, 117, 128, 176

Pomology, 6, 31-32

R

Rhizomes, 21, 95, 140, 181

S

Sapindaceae, 168

Sclerenchyma, 33, 47-48, 54, 60, 66, 69-71, 79, 81, 114, 162

Seed Germination, 77, 85, 161

Shoot System, 95-96, 169

Sophora Secundiflora, 14

Spongy Mesophyll, 38, 50, 67, 111

Stomata Function, 85, 89

Syzygium Aromaticum, 85

T

Tannins, 101, 115, 158, 183

Tap Root System, 160

Trichomes, 65-66, 110-112, 122, 180

Turgor Pressure, 40, 44, 94, 100

V

Vascular Bundle, 38, 53, 57, 113

Vascular Tissue, 29, 36, 48-50, 64, 81, 87, 102, 111, 113-114, 162-163, 165, 168, 179, 181

W

Woody Dicots, 180-181

X

Xylem, 29, 36-38, 47-51, 53-61, 63-65, 70-72, 79, 99, 110, 113, 140-141, 162-163, 168, 177, 179-181

Z

Zingiber Officinale, 85

www.ingramcontent.com/pod-product-compliance
Lightning Source LLC
Chambersburg PA
CBHW062005190326
41458CB00009B/2973